International Conference on Materials Science and Manufacturing Engineering (MSME 2019)

IOP Conference Series: Materials Science and Engineering
Volume 727

Guangzhou, China
7 - 9 November 2019

ISBN: 978-1-7138-2586-9
ISSN: 1757-8981

Printed from e-media with permission by:

Curran Associates, Inc.
57 Morehouse Lane
Red Hook, NY 12571

Some format issues inherent in the e-media version may also appear in this print version.

Institute of Physics
Dirac House, Temple Back
Bristol BS1 6BE UK

Phone: 44 1 17 929 7481
Fax: 44 1 17 920 0979

techtracking@iop.org

Additional copies of this publication are available from:

Curran Associates, Inc.
57 Morehouse Lane
Red Hook, NY 12571 USA
Phone: 845-758-0400
Fax: 845-758-2633
Email: curran@proceedings.com
Web: www.proceedings.com

TABLE OF CONTENTS

MATERIALS ENGINEERING

DESIGN AND MANUFACTURING

IOP Conf. Series: Materials Science and Engineering **727** (2020) 011001 doi:10.1088/1757-899X/727/1/011001

PREFACE

2019 International Conference on Materials Science and Manufacturing Engineering (MSME 2019) was held in Guangzhou, China from Nov.7 to 9, 2019. MSME 2019 was organized by Hong Kong Society of Mechanical Engineers (HKSME), supported by South China university of technology. The conference provides a useful and wide platform both for display the latest research and for exchange of research results and thoughts in Materials Science, Manufacturing Engineering and other topics. The participants of the conference were from almost every part of the world, with background of either academia or industry, even well-known enterprise. The success and prosperity of the conference is reflected high level of the papers received.

The proceedings are a compilation of the accepted papers and represent an interesting outcome of the conference. This book covers 3 chapters: Materials engineering, Design and manufacturing and Aerospace engineering.

We would like to acknowledge all of those who supported MSME 2019. Each individual and institutional help were very important for the success of this conference. Especially we would like to thank the organizing committee for their valuable advices in the organization and helpful peer review of the papers.

We sincerely hope that MSME 2019 will be a forum for excellent discussions that will put forward new ideas and promote collaborative researches. We are sure that the proceedings will serve as an important research source of references and the knowledge, which will lead to not only scientific and engineering progress but also other new products and processes.

Prof. Ruxu Du, South China University of Technology, China

IOP Conf. Series: Materials Science and Engineering **727** (2020) 011002 doi:10.1088/1757-899X/727/1/011002

COMMITTEES

Conference Chairman

Prof. Ruxu Du, South China University of Technology, China

Program Committee

Prof. Hamid Assadi, Brunel University London, UK

Prof. Chen Zhan, Auckland University of Technology, New Zealand

Academic Committee

Dr. Ramesh Kuppuswamy, University of Cape Town, South Africa

Prof. Stefan Dimov, University of Birmingham, UK

International Technical Commmitte

Assoc. Prof. Ţălu Ştefan, The Technical University of Cluj-Napoca, Romania

Assoc. Prof. Xiaolong Chen, Jinan University, China

Dr. Han Zhang, Queen Mary University of London, UK

Dr. Dmitry Ivanov, University of Bristol, UK

Dr. Haiming Wen, Missouri University of Science and Technology, USA

Dr. Abhijit Pramanick, City University of Hong Kong, Hong Kong

Dr. Yulin Li, East China University of Science and Technology, China

Dr. Sam Hsien-Yi HSU, City University of Hong Kong, Hong Kong

Dr. Wan Sharuzi Wan Harun, Universiti Malaysia Pahang, Malaysia

Dr. Ario Sunar Baskoro, University of Indonesia, Indonesia

Dr. GUELAILIA Ahmed, Algerian Space Agency, ALGERIA

Peer review statement

All papers published in this volume of *IOP Conference Series: Materials Science and Engineering* have been peer reviewed through processes administered by the proceedings Editors. Reviews were conducted by expert referees to the professional and scientific standards expected of a proceedings journal published by IOP Publishing.

IOP Conf. Series: Materials Science and Engineering **727** (2020) 012001 doi:10.1088/1757-899X/727/1/012001

Mechanical Properties and Deformation Behaviour of ARCH and BCT Lattice Structures Manufactured by Selective Laser Melting

Ruochen DING[1,2], Baorui DU[1,*], Jun YAO[1], Huilong ZHENG[1], Yumeng Guo [1,2] and Zhenya KANG[1]

[1] Institute of Engineering Thermophysics, Chinese Academy of Sciences, Beijing 100190, China
[2] University of Chinese Academy of Sciences, Beijing 100049, China

*Corresponding author's e-mail address: dubaorui@iet.cn

Abstract. In this study, a new lattice structure ARCH is proposed, ARCH lattice structure and traditional lattice structures BCT were fabricated from 316 L stainless steel by SLM. Mechanical properties and deformation behavior of the lattice structures were experimentally investigated. The results show that the ARCH lattice structures have better mechanical property and energy absorption capability than BCT lattice structures. Under the same relative density, the compressive strength and elastic modulus of the ARCH lattice structure is 121.27% and 60.48% higher than the BCT lattice structure, and the ARCH lattice structures have better energy absorption properties than the BCT lattice structures.

Key words: lattice structure; selective laser melting; mechanical properties; energy absorption properties

1. Introduction

Lattice structures are porous structure composed of repeating cellular unit, which widely developed and used in aerospace, automobile, biomedicine and other fields because of their low density, high specific strength, high specific stiffness, good energy absorption ability, excellent thermal and acoustic insulation properties, etc. [1] However, the lattice structure is difficult to be manufactured with traditional manufacturing process due to its highly complex spatial structure. Additive manufacturing (AM) technologies are very suitable for making lattice structures. As a kind of the AM technologies, selective laser melting (SLM) have been successfully used to fabricate complex lattice structures for various purpose. The body-centered cubic (BCC), body-centered tetragonal (BCT), face-centered cubic (FCC), gyroid cellular (GC), rhombic dodecahedron (RD) and the structure derived from them were widely studied.

Shen et al. [2] tested three kinds of BCC lattice structures, and the results show that the SLM technique offers a viable method for manufacturing high-performance lattice structures. Long et al. [3] proposed a BCT lattice structure by removing the constraint of isotropy size in BCC lattice structure. The experimental result show that comparing to BCC structure, the relative density of BCT optimized structure increase only by 3.7%, but the elastic modulus and yield strength increase by 121.5% and 77.3%. Hao et al. [4] evaluates the manufacturability and performance of SLM produced GC lattice structures. The results reveal that the GC lattice structures can be manufactured free of defects by the SLM process without the need of support structures with a wide unit cell size range of 2 to 8 mm, the struts within the lattice structures with smaller unit cell sizes have higher densities and relative densities and the yield strengths and Young's module both decrease with the increase in the unit cell size of the lattice structures. Xiao et al. [5] experimentally studied the load bearing capacity of the RD lattice structures, the results showed the similar failure mode in both quasi-static and dynamic loadings, and the shear failure along the 45° shear band. Cao et al. [6] proposed a modified RD lattice structure, compare with the original RD lattice structure, the compressive modulus and initial yield strength can increase 79% and 55% respectively.

In previous research little attention has been focused on the arch strut lattice structures. In this paper, a new arch strut lattice structure is proposed. The aim of this paper is to compare the mechanical properties and deformation behaviour of arch strut and straight strut lattice structures.

2. Lattice geometry

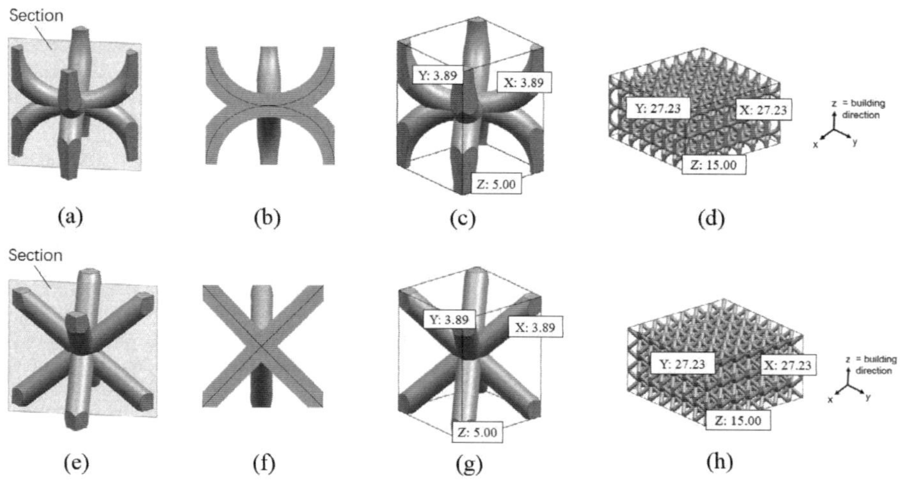

Figure 1. The CAD model of lattice structures

The geometry of the arch strut and straight strut unit-cell is given in Figure 1. (a) and (e). The arch strut unit-cell was named ARCH and the straight strut unit-cell is BCT. The centreline of ARCH is arched (the ratio of height and span of the arch is 0.45), as shown in Figure 1. (b), it is well known that compare with other structures, arch can afford bigger pressure. Dimensions of the lattice unit cells are defined in Figure 1. (c) and (g), arch strut of 1 mm and straight strut of 0.92 mm in diameter at the same relative density 0.2 has been selected for the experiments. The samples for mechanical testing are constructed. The CAD model of ARCH lattice structure and BCT lattice structure for compression testing are shown in Figure 1.(d) and (h) by merging 7 unit cells in x-, y- and 3 unit cells in z-directions.

3. Experimental details

3.1. Material and Fabrication

The lattice structure samples were made from 316L stainless steel powder which was produced by LPW Technology Ltd., UK. The particle size distribution of the 316L stainless steel powder is 10 ~ 45 μm. The ARCH and BCT lattice structure sample were fabricated by SLM using the LASERTEC 30 SLM printer from DMG with the material 316 L stainless steel. The laser power was 245 W, the layer thickness was 50 μm and the laser scanning speed was 1 m/s. Thermal treatment was applied at 1150 °C for 2 hours to reduce the residual stress. The lattice structure samples are displayed in Figure

IOP Conf. Series: Materials Science and Engineering **727** (2020) 012001 doi:10.1088/1757-899X/727/1/012001

2, three samples of each structure were fabricated to carry out the repeated experiments.

3.2. Tests

The lattice structure sample static compression testing was carried out on an INSTRON 5984 testing machine at room temperature under a uniaxial loading strain rate of 0.05mm/min. A digital camera was used to record the compress deformation of the samples.

Figure 2. The lathed uniaxial tensile samples and the lattice structure samples

4. Results and discussion

4.1. Compression properties

The quasi-static compression stress-strain curves of ARCH and BCT are shown in Figure 3. At the beginning of the compression, the stress-strain curves are non-linear and concave. The causes of this phenomenon maybe the uneven upper surface of the lattice sample. With the process of compression, the stress-strain curves entered a linearity stage, then the lattice start to damage and the stress-strain curves entered a long yield plateau. Finally, the entire structure collapses, struts are pressed together and densified, the stress-strain curves rise sharply. The compressive strengths and elastic modulus of ARCH and BCT lattice structures samples can be obtain from the stress-strain curves. Because of there is no clear first peak on the stress-strain curves, the arithmetical mean of the stresses between 20% and 40% of compressive strain is taken as the compressive strength of the lattice structure samples. The compressive strength of the ARCH lattice structure samples is 63.57 MPa, and the compressive strength of the BCT lattice structure samples is 28.73 MPa. The compressive strength of the ARCH lattice structure is 121.27% higher than the BCT lattice structure at the same relative density of 20%. The elastic modulus determined by gradient of the elastic straight lines determined by elastic loading and unloading between stresses of 70% of compressive strength and 20% compressive strength. The elastic modulus of the ARCH lattice structure samples is 3.37 MPa, and the elastic modulus of the BCT lattice structure samples is 2.10 MPa. The elastic modulus of the ARCH lattice structure is 60.48% higher than the BCT lattice structure at the same relative density of 20%.

Figure 4 shows the quasi-static compression deformation behaviour of ARCH and BCT lattice structures samples under increasing strain. The main failure mode of ARCH lattice structure samples is diagonal direction inclined compacted collapse, and the BCT lattice structure sample is vertical direction compacted collapse.

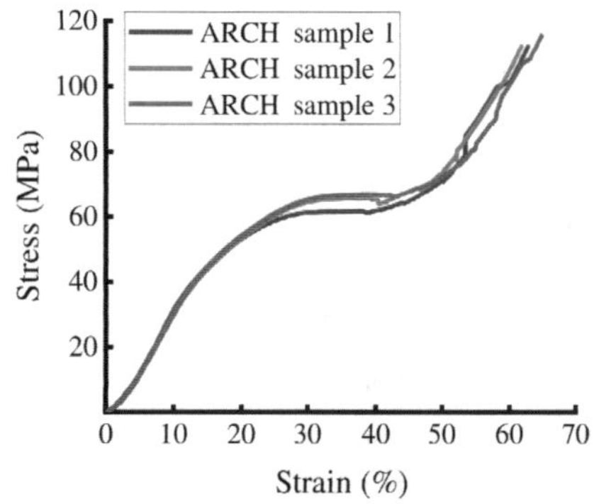

(a) ARCH

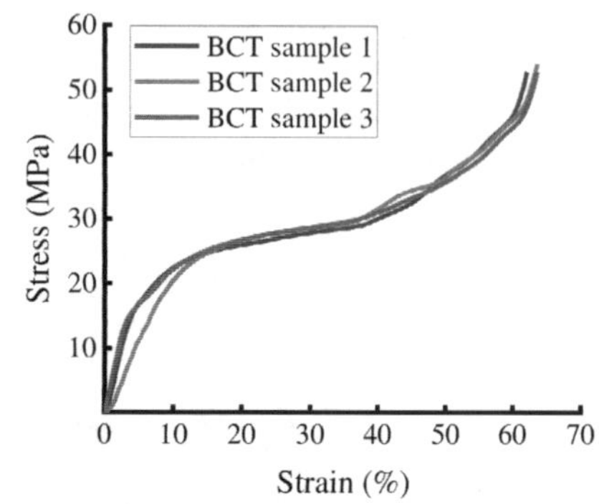

(b) BCT

Figure 3. The quasi-static compression stress-strain curves lattice structures samples

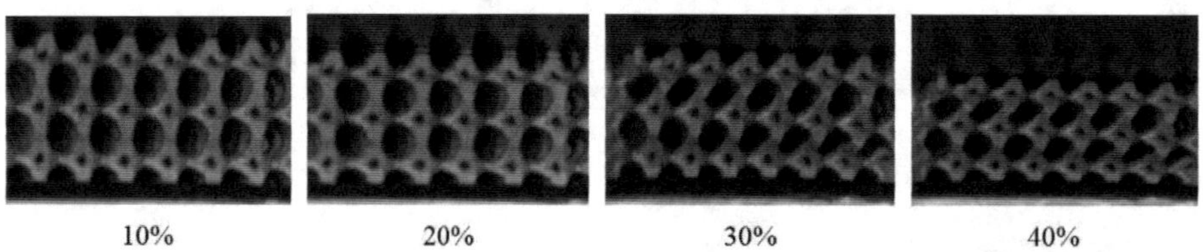

(a) ARCH

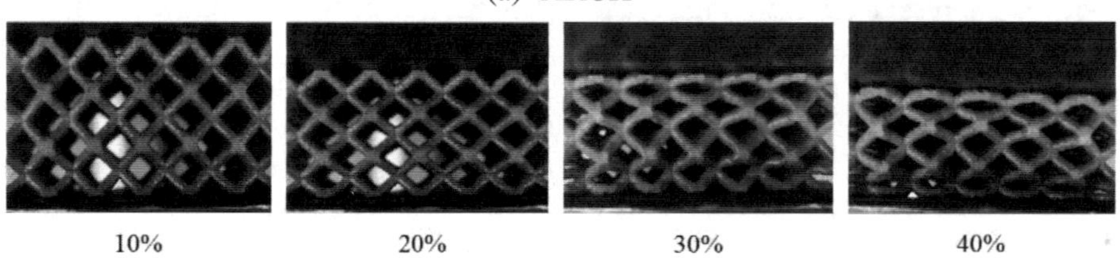

(b) BCT

Figure 4. The quasi-static compression deformation behaviour of lattice structures sample

4.2. Energy absorption of lattice structures

The specific energy absorption (SEA) and energy absorption efficiency are two important indicators to measure the impact resistance and energy absorption properties of lattice structures [1, 7]. The definition of SEA is namely absorbed energy per unit mass, and the SEA can be calculated by:

$$SEA = \frac{\int_0^{\varepsilon} \sigma(\varepsilon) d\varepsilon}{\rho_r} \tag{1}$$

Where, σ is the stress of the lattice structures [MPa]; ε is the strain of the lattice structures; ρr is the relative density of the lattice structures.

The energy absorption efficiency (η) is the ratio of the absorbed energy to the corresponding compression stress and the η can be calculated by:

$$\eta = \frac{\int_0^{\varepsilon} \sigma(\varepsilon)\, d\varepsilon}{\sigma} \tag{2}$$

The SEA and energy absorption efficiency curve were obtained by calculating the experiment data as shown in Figure 5. The SEA of ARCH and BCT lattice structures increase with the increase of the strain. The SEA of ARCH and BCT is very close and the SEA of BCT is a little bit higher than ARCH when the strain of the samples is less than 12.3%. While when the strain is bigger than 12.3% the SEA of ARCH is increase rapidly and much higher than BCT with the increase of strain. The SEA of

ARCH is 90.1% higher than BCT when the strain is 50%. The energy absorption efficiency of ARCH and BCT is firstly increased and then decreased with strain increase. When the strain is less than 10.7%, the energy absorption efficiency of ARCH and BCT are basically the same, with the increase of strain the energy absorption efficiency of ARCH is higher than BCT and the difference between the two structures are getting bigger. The energy absorption efficiency of the ARCH and BCT is reaches the maximum at the strain is 50 % and 61 % respectively, and the maximum energy absorption efficiency of the ARCH is 39.8% higher than BCT. Based on the analysis of the Figure 5, the ARCH lattice structures have better energy absorption properties than the BCT lattice structures.

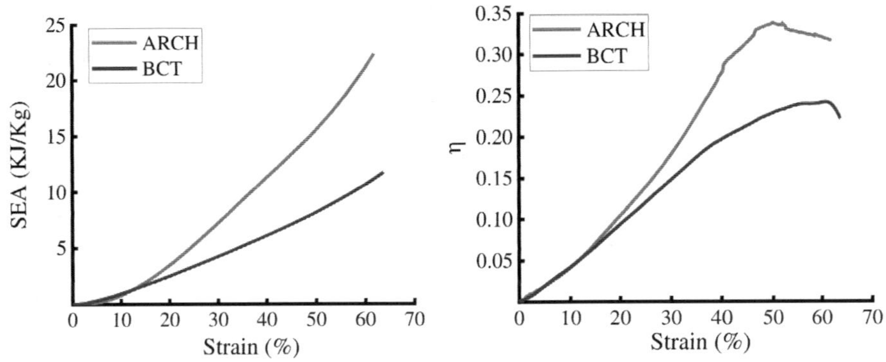

(a) SEA (b) Energy absorption efficiency

Figure 5. The energy absorption properties of lattice structures

5. Conclusion

In this paper, a new lattice structure ARCH is proposed, ARCH lattice structure and traditional lattice structures BCT were fabricated by SLM. Mechanical properties and deformation behaviour of the lattice structures were numerically and experimentally investigated were carried out to compare with traditional lattice structure. The following conclusions can be drawn from the present study:

- Under the same relative density, the compressive strength of the ARCH lattice structure is 121.27% higher than the BCT lattice structure and the elastic modulus of the ARCH lattice structure is 60.48% higher than the BCT lattice structure. The ARCH lattice structure has better mechanical properties than the BCT lattice structure.
- In the compression process, the lattice structure deformation is stable and continuous, there is no signs of local brittleness failure. The main failure mode of ARCH lattice structure samples is 45° inclined collapse, and the BCT lattice structure sample is compacted layer by layer.
- The SEA and energy absorption efficiency of ARCH and BCT with little difference at the beginning, then the SEA and energy absorption efficiency of ARCH increase rapidly and much higher than BCT. The ARCH lattice structures have better energy absorption properties than the BCT lattice structures.

References

[1] Gibson LJ, Ashby MF. Cellular solids: structure and properties. *Cambridge University Press*, 1997.

[2] Shen Y, Mckown S, Tsopanos S, Sutcliffe C, Mines R and Cantwell W. The Mechanical Properties of Sandwich Structures Based on Metal Lattice Architectures. Journal of Sandwich Structures and Materials 2010, 12(2):159-180.

[3] Long B, Fei X, Xiaohong C, Changyan Y, Junfang Z, Rui C. Multi-objective Structural Optimization Design of Ti6Al4V Lattice Structure Formed by SLM.. *Journal of Mechanical Engineering*, 2018, 54(5): 156-165.

[4] Hao L, Raymont D, Yan C, Hussein A, Young P, Evaluations of cellular lattice structures manufactured using selective laser melting, *International Journal of Machine Tools & Manufacture*, 2012,62(11):32-38.

[5] Babaee S , Jahromi B H , Ajdari A , et al. Mechanical properties of open-cell rhombic dodecahedron cellular structures. *Acta Materialia*, 2012, 60(s 6-7):2873-2885.
[6] Xiao L, Song W, Wang C, Liu H, Tang H, Wang J. Mechanical behavior of open-cell rhombic dodecahedron Ti–6Al–4V lattice structure. *Materials Science & Engineering A*, 2015, 640(7):375–84.
[7] Qingming L, Magkiriadis I, Harrigan J. Compressive Strain at the Onset of Densification of Cellular Solids. *Journal of Cellular Plastics*, 2006, 42(5):371-392.

IOP Conf. Series: Materials Science and Engineering **727** (2020) 012002 doi:10.1088/1757-899X/727/1/012002

Comparison of Sliding Wear Properties of 400 Grade Low Alloy Wear-resistant Steels with Different Microstructure

YingChao Pei[1], DianXiu Xia[1,*], Liang Cong[2], XueLin Wang[3], ShouRen Wang [1] and DongYue Wang[1]

[1] School of Mechanical Engineering, University of Jinan, Jinan 250022, China
[2] Bisalloy Shangang (Shandong) Steel Plate Co., Ltd, Jinan 250022, China
[3] Collaborative Innovation Center of Steel Technology, University of Science and Technology Beijing, Beijing 100000, China

[*]Corresponding author's e-mail: xdxlxf@126.com

Abstract. Comparative study on sliding friction and wear properties of low alloy wear-resistant steel Hardox400 and NM400 at room temperature using Retc multi-function friction and wear tester. The hardness, microstructure and wear surface morphology of the test steel were compared and analyzed by microhardness tester, optical microscope (OM) and scanning electron microscope (SEM). The results show that the wear mechanism of the two kinds of test steels is mainly manifested as abrasive wear and fatigue wear. The Martensite-Bainite Dual Phase Structure in Hardox400 has a good match of strength and toughness, and the sliding wear performance is better than the NM400 of martensite structure.

Key works: sliding wear; microstructure; wear properties

1. Introduction

The three most common failure modes of metallic materials are wear, corrosion and fatigue[1-2]. According to incomplete statistics, about 30% of China's energy consumption is caused by different forms of wear and tear every year, which not only leads to serious consumption of energy and resources, but also causes huge losses to the national economy[3-4]. In recent years, low-alloy high-strength wear-resistant steel has been widely used in the manufacturing of mechanical equipment in mining, coal and transportation industries due to its low alloy content, low economic cost and excellent comprehensive performance[5-6]. However, due to the relatively late start of the research and development of low-alloy wear-resistant steel in China, the wear properties and stability of wear-resistant products in China still have a certain gap compared with the same grade products in foreign countries[7-8]. This article to a domestic steel production NM400 wear-resisting products and Sweden Hardox400 wear-resisting products manufactured by an enterprise as the research object, under the same experimental conditions of two kinds of wear resistant steel reciprocating sliding wear experiment. The reasons for the difference of wear resistance between the two kinds of test steels are briefly discussed from the point of view of microstructure. It provides theoretical data support for the research and development of domestic low alloy wear-resistant steel products.

2. Experimental materials and methods

The experimental steel is Hardox400 steel plate produced by Sweden and NM400 steel plate produced by a domestic steel mill, with the thickness of 50mm. The chemical composition of the two experimental steel plates is shown in table 1.

Table 1. Chemical composition of testing steels(mass fraction, %)

Steel	C	Si	Mn	P	S	Cr	Mo	B	Ti
Hardox400	0.12	0.49	1.28	0.01	0.002	0.65	0.02	0.001	0.02
NM400	0.17	0.30	1.30	0.01	0.002	0.35	0.20	0.001	0.02

The mechanical properties of the two experimental steels are shown in table 2.

Table 2. Mechanical properties of testing steels

Steel	$R_{P0.2}$/MPa	R_m/MPa	A_{KV}/J(-20°C)
Hardox400	1080	1220	50
NM400	1130	1250	30

In this study, the test steel surface is selected as the surface to be tested, and the test steel surface shall be removed with a thickness of 0.3mm to avoid the influence of surface decarburization layer on the experimental results. The sliding wear test was carried out on Retc multifunctional friction wear test machine. The size of the sample is 10mm×10mm×5mm. Before the sliding wear test, the surface to be tested should be ground and polished. The grinding ball is made of GCr15 bearing with a diameter of 6.5mm and a hardness of 750HV. Experimental parameters are set as follows: Load 50N, frequency 4HZ, amplitude 4.5mm, experimental time 45min. The experimental principle is shown in figure 1. The friction coefficient, wear volume, depth and morphology of the test steel were recorded. Wear morphology of experimental steel was observed by JSM-7610F scanning electron microscope. The surface hardness were measured by Wilson 402MVD microhardness tester. The size of the sample is 10mm×10mm×10mm. After the hardness test, the sample was ground and polished for corrosion. The corrosive agent is a 4% nitrate alcohol solution by volume fraction. The microstructure of the steel was observed by using VHX-2000C optical microscope and JSM-7610F scanning electron microscope. After that, the sample was ground, polished, corroded and placed under an optical microscope to observe the original austenite structure. The corrosive agent is a mixture of distilled H_20 + CCl_4 + $C_6H_2OH(NO_2)_3$ + shampoo + HCl.

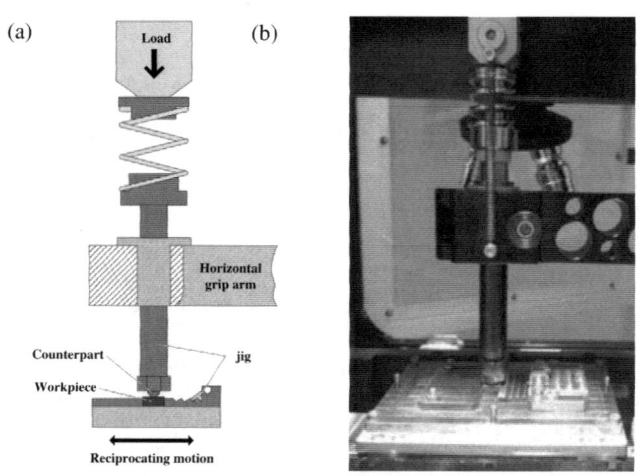

Figure 1. Principle diagram of friction and wear (a) Schematic diagram,(b) Equipment diagram

3. Experimental result

3.1. Sliding wear performance comparison

3.1.1. Coefficient of sliding friction

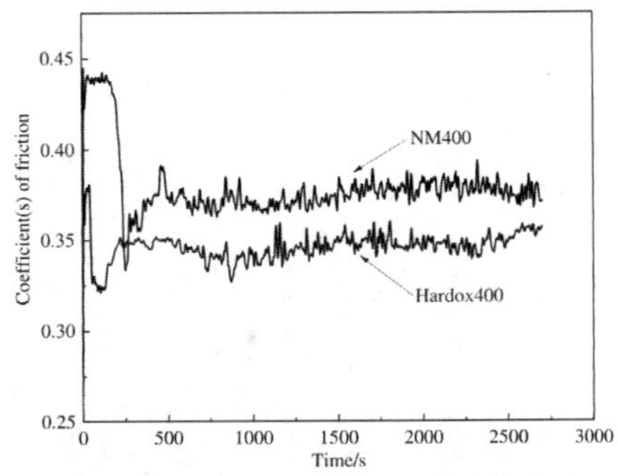

Figure 2. Friction coefficient with time

Figure 2 shows the experimental results of friction coefficients of test steels. It can be seen that the friction coefficients of the two kinds of test steels have the same trend with time. They all rise first, then fall and then tend to be stable. But there are some differences in the friction coefficients. The average friction coefficient of Hardox400 is 0.3463, and NM400 is 0.3787.

3.1.2. Wear resistance contrast

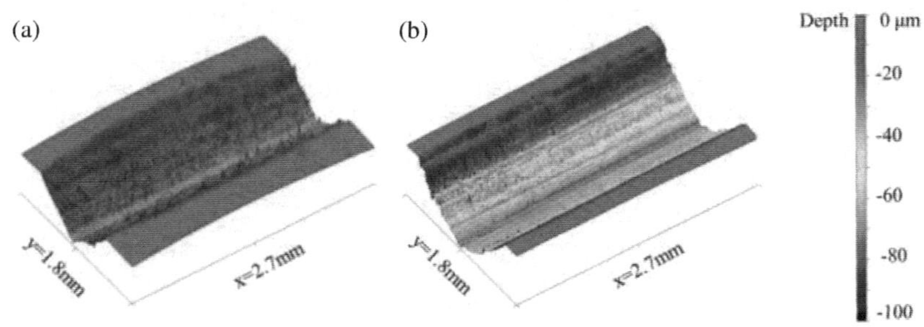

Figure 3. The macroscopic 3D morphology of sliding wear of testing steels (a)Hardox400, (b) NM400

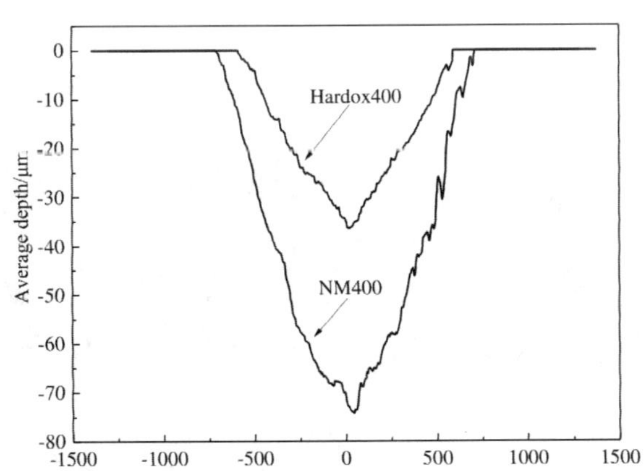

Figure 4. Contour curve of the cross section at the middle position of the wear mark of testing steels

Table 3. Average wear depth, width and wear volume of testing steels

Steel	Average wear depth /µm	Average wear width /µm	wear volume /mm^3
Hardox400	35.8	108.4	0.2294
NM400	75.3	148.6	0.3704

Figure 3 shows the macroscopic wear 3D morphologies of test steels. The cross section profile sizes of the two test steel wear marks were further measured, and the corresponding contour curves were drawn by taking the average value, as shown in Figure 4. Table 3 shows the experimental results of average abrasion mark depth, abrasion mark width and abrasion volume of test steels obtained through measurement and calculation. By comparison, it is found that the width, depth and wear volume of NM400 are all larger than Hardox400. The wear resistance of a material can usually be evaluated by the reciprocal of the wear rate[9]. It can be concluded that the abrasion resistance of Hardox400 experimental steel is 196.1639, while that of NM400 experimental steel is only 121.4903.

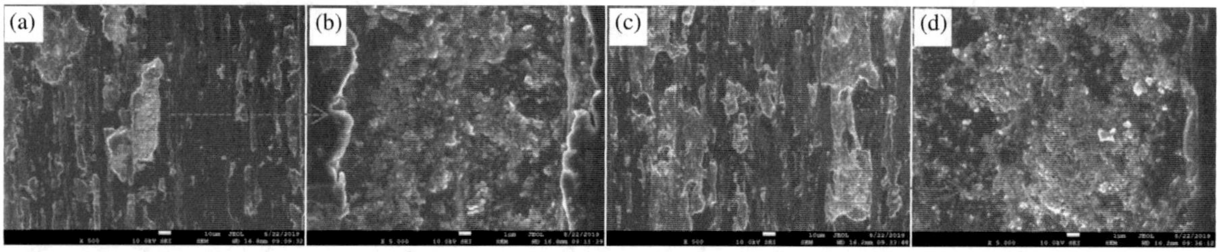

Figure 5. The microstructure of sliding wear of testing steels (a,b) Hardox400,(c,d) NM400

The microabrasion morphologies of the two test steels under SEM are shown in Figure 5. It can be seen that the wear surfaces of the two kinds of experimental steels are mainly characterized by extrusion deformation and fatigue spalling. The wear mechanism is different degrees of abrasive wear and fatigue wear. By comparing Figure 5 (a) and (c), it can be seen that the fatigue spalling of NM400 is more serious than that of Hardox400. The NM400 fatigue spalling pit has more non-spalling abrasive particles, which are more densely distributed and easier to spalling in the wear process.

3.2. Hardness

Table 4. The surface hardness of steel/HV

Steel	1	2	3	4	5	Average
Hardox400	410.71	402.37	405.94	414.98	411.97	409.20
NM400	403.08	410.08	418.64	407.40	406.05	409.05

Table 4 shows the hardness measurement results. The surface hardness of Hardox400 was 409.20HV, NM400 was 409.05HV. It can be seen that the hardness of the two test steels is similar.

3.3. Microstructural

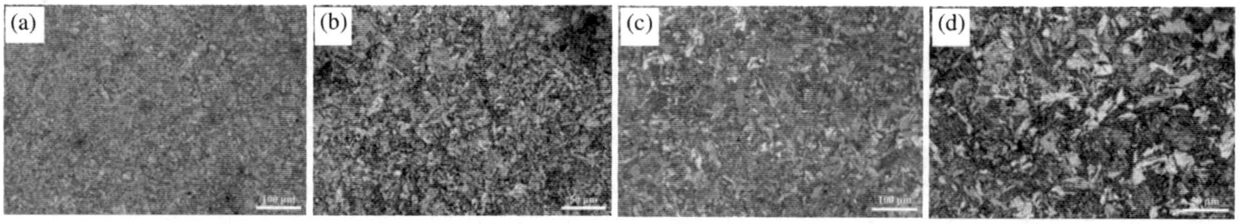

Figure 6. optical microstructure of steel surface(a,b) Hardox400,(c,d) NM400

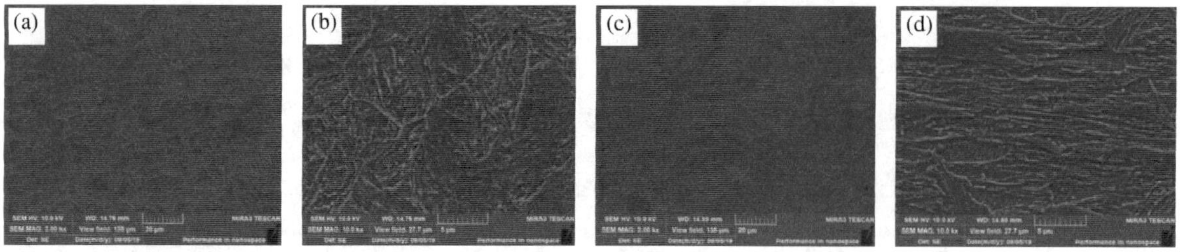

Figure 7. SEM microstructure of steel surface (a,b) Hardox400,(c,d) NM400

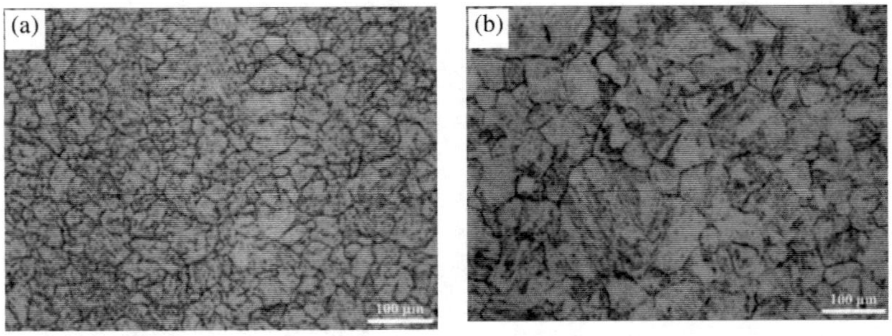

Figure 8. Austenite morphology of steel surface (a) Hardox400, (b) NM400

The microstructure of the test steel are shown in figure 6 and figure 7. Under an OM, NM400 is a typical lath martensite. The microstructure of Hardox400 was different from that of NM400, but the grains were small and dense.It can be found by SEM that there are many martensite lath bundles with different orientation in NM400 original austenite grain, and small carbides are scattered in and between the martensite lath. Hardox400 is Martensite-Bainite Dual Phase Structure. The martensite lath in the original austenite grain is shorter, the martensite lath bundle is more densely distributed, and a small amount of short and narrow lower bainite structure remains between the martensite lath. Figure 8 shows the original austenitic shape of the test steel. It can be seen that most grains of Hardox400 are relatively small and evenly distributed. However, most of the original austenite grains in NM400 were relatively large, and the coarse original austenite grains were surrounded by a small number of fine original austenite grains, and the grain size difference was obvious.

4. Discuss the analysis

Combined with the above test results, it is concluded that the sliding wear performance of Hardox400 with Martensite-Bainite Dual Phase Structure under the experimental conditions is better than that of NM400 with lath martensite structure. According to the hardness test results, the hardness of Hardox400 is similar to that of NM400. Analysis shows that, the carbon content of lower bainite in the complex phase of Hardox400 is low, and most of the carbon elements in the tissue are involved in the formation of martensite, which makes the martensite structure formed higher hardness. And the original austenite grain size of Hardox400 is smaller and the grain boundary density in the large-angle structure is higher, which makes the matrix structure obtain higher hardness[10]. It has been shown that when the hardness of the material is similar, the wear-resisting performance of the material is mainly affected by the toughness of the matrix[11-13]. The lower bainite structure in the complex phase of Hardox400 is distributed between lath martensite, which can effectively inhibit the crack growth, making the toughness of Hardox400 in the complex phase of Hardox400 is better than that of NM400 in the martensite. The Hardox400 tissue is relatively small and evenly distributed, which increases the grain boundary area in the tissue, and makes the crack growth require more energy, reduces the crack growth rate in the matrix tissue, and improves the toughness of the matrix tissue[14-15]. Therefore, Hardox400 has less fatigue spalling and smaller wear volume in the wear process. The structure of NM400 is lath martensite, which has a high content of supersaturated carbon, high hardness but poor toughness. In the process of wear, it is easy to lead to the initiation and expansion of cracks, thus causing serious fatigue spallation. The wear mechanism of the two kinds of test steels shows fatigue wear and abrasive wear. However, due to the relatively poor toughness of the martensite matrix of NM400, the fatigue spalling of the worn surface is more serious, resulting in the sliding wear performance of NM400 is not as good as that of Hardox400.

5. Conclusion

(1)Under the same hardness, the wear-resistant properties of wear-resistant steel depend on the microstructure of the material. Under the test conditions, the sliding wear performance of Hardox400 wear-resistant steel with Martensite-Bainite Dual Phase Structure is better than that of NM400 wear-resistant steel with martensite matrix structure.

(2)Under the conditions of loading of 50N, frequency of 4HZ, amplitude of 4.5mm, and wear time of 45min, the wear mechanism of the two kinds of experimental steels shows fatigue wear and abrasive wear. Furrows plastic deformation and fatigue spalling occurred on the wear surface.

References

[1] Li Wenbin, Fei Jing, et al. (2012) Production status and development direction of wear resistant steel with low alloy and high strength in China[J]. Materials for Mechanical Engineering, 36(2):6-10.
[2] Deng Feng, Hu Feng, Wu Kaiming. (2016) Development and application of low-alloy high-strength wera-resistant steel[J]. Metal Materials and Metallurgy Engineering,44(2):29-35.
[3] Zhang Yubin, Qin Jie. (2009) Current status and development of high strength wear-resistant steel production[J]. World Iron and Steel, (6): 23-26.
[4] Ma Heng, Zhou Ping, Liu Jungang, et al. (2011) Effects of tempering temperature on microstructure and properties of NM360 wear-resistant steel[J]. Heat Treatment of Metals, 36(4):66-68.
[5] Lemm J D, Warmuth A R, Pearson S R, et al. (2015) The influence of surface hardness on the fretting wear of steel pairs--Its role in debris retention in the contact[J]. Tribology International, 81:258-266.
[6] José Rendón, Olsson M. (2009) Abrasive wear resistance of some commercial abrasion resistant steels evaluated by laboratory test methods[J]. Wear, 267(11):2055-2061.
[7] Autay R, Kchaou M, Dammak F. (2012) Friction and Wear Behavior of Steels under Different Reciprocating Sliding Conditions[J]. Tribology Transactions,55(5):590-598.
[8] Ojala N, Valtonen K, Antikainen A,et al. (2016) Wear performance of quenched wear resistant steels in abrasive slurry erosion[J].Wear,354-355:21-31.
[9] Jia-jun L. (1993)Principle of material wear and its wear resistance [M]. Tsinghua University Press.
[10] Binbin W , Zhiquan W , Xuelin W, et al. (2019)Relationship between high angle grain boundaries and hardness after $\gamma\rightarrow\alpha$ transformation[J]. Materials Science and Technology,vol.35 ,NO.15,1803-1814
[11] Gola A M, Ghadamgahi M, Oi S W. (2017) Microstructure evolution of carbide -free bainitic steels under abrasive wear conditions[J].Wear,376-377:975-982.
[12] Trevisiol C,Jourani A,Bouvier S. (2018) Effect of microstructures with the same chemical composition and similar hardness levels on tribological behavior of a low alloy steel[J].Tribology International.
[13] Ojala N, Valtonen K, Heino V, et al. (2014) Effects of composition and microstructure on the abrasive wear performance of quenched wear resistant steels[J]. Wear,317(1-2):225-232.
[14] Guo-Ning H, Ning-Bo Z, Chao-Lei Z, et al. (2018) Effect of Niobium on the Dry Sliding Friction and Wear Properties of 100Cr6 Bearing Steel[J]. Journal of Tribology,140(6).
[15] Valtonen K, Ojala N, Haiko A, et al. (2019) Comparison of various high-stress wear conditions and wear performance of martensitic steels[J]. Wear, (3-13):426-427.

IOP Conf. Series: Materials Science and Engineering **727** (2020) 012003 doi:10.1088/1757-899X/727/1/012003

Effect of Surface State on Acid Rain Corrosion Resistance of T6 6005A Aluminum Alloy by BT-FSW Joint

Zhiheng Qin* and Hongji Xu

School of Materials Science and Engineering, Dalian Jiaotong University, Dalian, Liaoning Province, 116028, China

*Corresponding author: E-mail: 494905023@qq.com

Abstract. The test sample was 5 mm thick artificial aging 6005A aluminium alloy profile, which was butted by BT-FSW. This work aimed to analyse the effect of surface state on corrosion resistance of welded joints in acid rain environment based on acetate ssalt spray test (CASS), polarization curve and EIS impedance test, combined with SEM scanning and EDS energy spectrum. The results showed that the corrosion rate of pre-welded polished joints was high at the initial stage of salt spray corrosion, and the corrosion rate of both joints were equal at the later stage of corrosion, higher than that of base metal. The best preservation of corrosion morphology was THAZ, followed by nugget zone and HAZ, and the pitting degree of base metal was the most serious. The polarization curves of the polished joints before 30 min corrosion were located at the lower right of the image, and its self-corrosion current density in the nugget zone showed the worst corrosion resistance. The corrosion resistance of the pre-welded polished joint increased after 336 hours of corrosion, the corrosion breakdown potential and self-corrosion current were not significantly different from those of the unpolished joints. The fitting curves of the two kinds of joints had similar dual capacitance arc resistance characteristics and the impedance parameters in EIS, and the electrochemical corrosion resistance of the two joints were weaker than that of the base metal.

1. Instruction

6005A is Al-Mg-Si heat-treppable aluminum alloy, which has been applied in batches in the field of rail vehicles. Its production process has realized the effect of refining the grain by increasing the content of Cr and Mn elements, effectively inhibiting the occurrence of recrystallized grains during hot processing of extruded profiles[1]. After the introduction of the FSW technology, bringing about tremendous changes, and its joint performance of aluminum alloy has been greatly improved. Bobbin Tool Friction Stir Welding (BT-FSW) is used as a plastic joining method. It is widely used in the welding of aluminum alloy thin plates, hollow siding and structural profiles, as a supplement to the traditional friction welding. Therefore, the 6005A aluminum alloy is an ideal connecting material for the side wall of rail vehicles[2]. However, its application is limited by environmental factors such as humidity, temperature, pH, atmospheric acid ion concentration and strong oxidizing gas that make the corrosion law of the joint become elusive. There have been many available reports on the corrosion resistance of aluminum alloy in China. Shenghai et al[3] studied the effect of pH on the electrochemical properties of 2024-T351 aluminum alloy. It was found that the stability of the oxide film decreased with the decrease of pH. When pH below 3, the oxide film basically lost its ability to protect the substrate. Li Tao et al[4] studied the 2A12 aluminum alloy dynamic electrochemical impedance spectroscopy (EIS) by dry and wet weekly immersion test. The scanning results showed that increasing the concentration of Cl^- promoted the formation and development of pitting corrosion. The pre-welded polished treatment of traditional MIG welding is to remove the oil stain and oxide film on the surface of the sheet, avoiding the pores and inclusion defects of the joint. At present, for the 6 series aluminum alloy profiles BT-FSW joint, the different surface polishing state joints in the acid rain atmosphere corrosion behavior remains to be studied. In this paper, the corrosion behavior and mechanism were studied under environment of simulated acid rain on the artificially aged (T6) 6005A profiles' BT-FSW joints by static weight loss, salt spray test and potentiodynamic polarization curve.

2. Research methods

2.1. Test materials

The test material is 5 mm thick 6005A aluminum alloy profiles with T6 artificial aging. The pre-welded polishing adopts the method of thousand-blade grinding. The chemical composition is shown in Table 1, and welding parameter is shown in Table 2.

Table 1. Chemical composition of 6005A-T6(mass fraction, %)

Brand	Si	Fe	Cu	Mn	Mg	Cr	Zn	Ti	Al
6005A-T6	0.59	0.18	≤0.09	≤0.12	0.57	≤0.1	0.02	0.01	Bal.

Table 2. Process parameters of BT-FSW

Stirring needle size (mm)			Inclination (°)	Angle (°)	Friction speed (rpm)	Welding speed (mm/min)	Pressing amount (mm)
Needle	Upper	Lower					
4.3	23	20	0	0	400	300	0.7

2.2. Salt spray test

According to the standard GB/T 10125-2012, the KLT-2008 test chamber is used for acetate ssalt spray test (CASS test) to test the corrosion resistance of the joint specimen under the simulated industrial atmospheric acid rain corrosion environment. The test chamber temperature is set to 50°C, using continuous spray method. The concentration of 1 mol/L NaCl+0.26 mol/L $CuCl_2$ salt solution was added to the brine tank, and the pH value of the solution was adjusted to 3.0 by adding appropriate amount of glacial acetic acid. Each test piece' size was 70 mm×70 mm×5 mm and is weighed before welding. Before the corrosion test, the non-corrosive surface of the surrounding and back sides is sealed with epoxy resin. The test period contiances 24h, 48h, 72h, 96h, 120h, and 240h. Three parallel test pieces are selected for each cycle. After reaching the specified test period, the test pieces are recorded the weight after ultrasonically cleaned, epoxy resin degreased and anhydrous ethanol dehydrated. The weight loss rate and corrosion rate are calculated for drawing the curve of weight loss. Finally, the microscopic corrosion morphology of the specimen surface was observed by SUPRA 55 field emission scanning electron microscope.

2.3. Electrochemical test

According to the standard GB/T 24196-2009, two kinds of surface state joint potentials polarization curve (Tafel curve) and electrochemical impedance spectroscopy (EIS) were measured using CorrTest electrochemical workstation. 0.6 mol/L NaCl+0.01 mol/L $NaHSO_3$+0.01 mol/L $NaHCO_3$+0.01 mol/L $NaNO_3$ salt solution was used as the corrosion medium to simulate the acid rain environment with pH less than 4.0 at room temperature. The size of test sample was 10 mm×10 mm×5 mm. Before electrochemical test, polishing with metallographic sandpaper (1000 mesh) to removing the surface oxide film to be corroded to uniform roughness. The test corrosion cycle is immersion corrosion for 30 min short-term and 336 h long-term immersion corrosion, using a typical three electrode system (working electrode, SCE saturated calomel electrode and platinum plate assisted electrode), polarization ranges -2.0V~0V, scanning speed is 1.68mV/s. When measuring EIS impedance spectrum, the scanning frequency range selects 0.01 HZ~10^5 HZ, test potential selection the open circuit potential, the peak value of the sinusoidal disturbance signal is 5mV, and the experimental data is finally processed by Zsimpwin.

3. Test results and analysis

3.1. Salt spray test

After 240 hours salt spray test, the T6 state 6005A welded joints parent metal area was severely corroded, and the surface was evenly distributed densely. The corrosion production was stacked in a block shape. At this time, the corrosion pit had a tendency to develop into a large piece of peeling pit and could be observed obviously in the joint surface. Starting from the junction of THAZ and HAZ on both sides, forming a fine etch pit of the contiguous. The pit distribution had a linear distribution of the second phase, indicating that the second phase particle Mg_2Si was a sensitive position of the etch[5]. The corrosion of the retreating side of the THAZ was lighter than that of the advancing side. The number and depth of the pitting pits on Nugget zone of the two kinds of joints wre significantly smaller than those of parent metal and HAZ, indicating that the corrosion resistance of BT-FSW welded joints was higher than that of parent metal.

Each Surface State test piece takes 3 parallel test sample, and the average corrosion rate is calculated according to the following formula:

$$\upsilon = \frac{m_0 - m_1}{st} \qquad (1)$$

$$\bar{\upsilon} = \left(\sum_{i=1}^{i=3} \upsilon_i\right)\frac{1}{3} \qquad (2)$$

Where υ is the corrosion rate of a single test piece, m_0 is the original weight of the test piece, m_1 is the weight after corrosion of the test piece, s is the area of the corrosion surface, t is the corrosion time, $\bar{\upsilon}$ is the average corrosion rate of the piece. The results of parent metal and welded joints corrosion weight loss rate are shown in figure 1.

It can be seen from Figure 1(a) that the corrosion weight loss rate of parent metal was small and the pre-welded polished joint was the highest within 120 h of corrosion time. After corrosion for more than 120 h, the corrosion weight loss rate of parent metal increased rapidly, and that of two surface state welded joints were low and tended to be consistent. When the corrosion time reached 240 h, parent metal had the highest corrosion weight loss rate, reaching 111.84 g/m2.

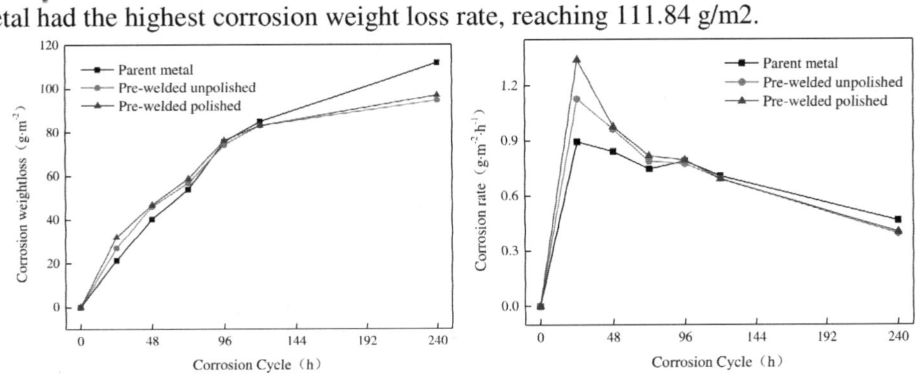

(a) corrosion weight loss curve (b) corrosion weight loss rate curve

Figure 1. Curves of corrosion weight loss and corrosion weight loss rate

It can be clearly seen from Figure 1(b) that the corrosion rate of the two welded joints was significantly higher than that of parent metal within 120 h at the beginning of the test. The corrosion rate of pre-welded polished joint was the highest, reaching 1.34 g/(m2·h), and the corrosion rate rised fastest in the initial 24 h. At this time, the active anions in the salt solution (such as Cl^-, SO_3^{2-}, HCO_3^-,

NO_3^-, etc.) was the main corrosion driving force. Corrosion from 24 h to 240 h, corrosion rate through the process of "reduction - increase - decrease". Because of gradually accumulating corrosion production layer, the effect steric protection hindered the contact of the active anion in the corrosive medium with the aluminum alloy matrix, which caused the corrosion rate to decrease in a short period of time. After 72 h, the corrosion rate rised again. It was presumed that the surface corrosion pit expanded along the grain boundary, and the internal expansion of the corrosion production inside the corrosion pit causes the large corrosion production attached to the surface to split, creating a new channel for the corrosive medium to contact the metal matrix. After 120 h of corrosion, the weight loss rates of the two surface state welded joints tended to be the same, indicating that the coverage of the joint surface corrosion production was stable status. It effectively blocked the active anion from contacting the metal surface. After 240 h, the weight loss rate of parent metal was higher, reaching 0.466 $g/(m^2 \cdot h)$. The change of corrosion rate showed that the pre-welded polished joint had poor corrosion resistance at the initial stage of the test. In the latter part of the test, the corrosion resistance of the two surface state welded joints was equivalent and better than parent metal.

In the CASS salt spray test, $CuCl_2$ salt solution can induce corrosion quickly. Figure 2 shows the SEM results of BT-FSW joint under simulated acid rain corrosion environment. It can be seen that after 120 h of corrosion, parent metal surface was covered with reddish brown loose pits, caused by the low Cu potential in an acidic environment. The copper-rich S phase (Al_2CuMg) and the θ phase ($CuAl_2$) were preferentially dissolved as anodes[6]. The red-brown corrosion production Cu_2O corrosion production growed up and cracked due to internal stress, forming new corrosion channels to aggravate pitting corrosion. Compared with parent metal, the corrosion phenomenon of Nugget was mainly shallow exfoliation, there were micro-cracks in the pitting area. THAZ had the most complete surface morphology, shallow pitting pits, and less corrosion production attached. HAZ substrate had a large number of small diameter corrosion production, and its corrosion pit was flaky, with a tendency to develop internally[7].

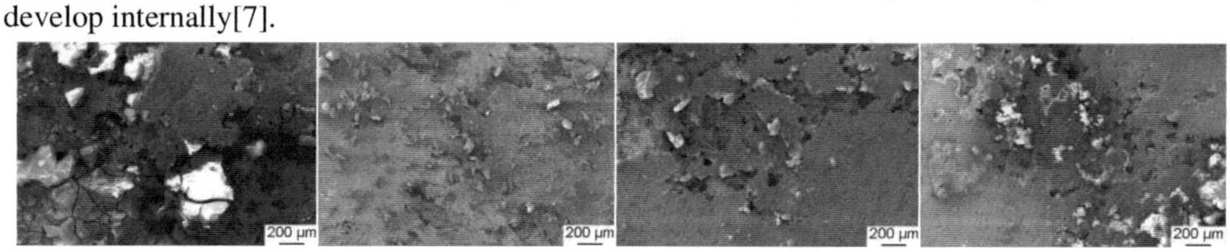

(a)Parent metal 150× (b)Nugget 150× (c)THAZ 150× (d)HAZ 150×

Figure 2. SEM images of welded joints after salt-spray corrosion

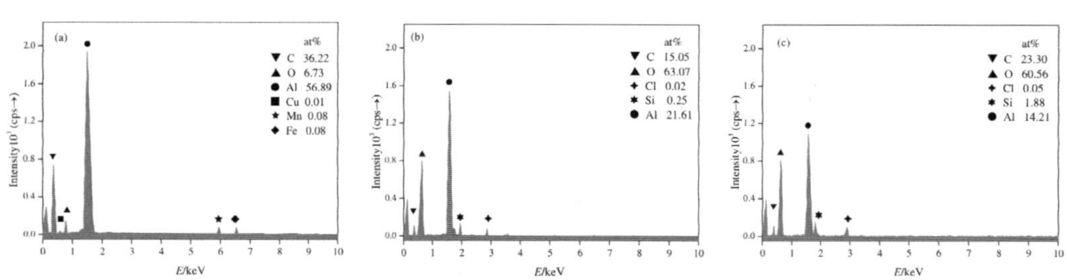

(a) Parent metal (b) Pre-weld unpolished joints (c) Pre-welded polished connector

Figure 3. EDS patterns of corrosion production

The EDS energy spectrum of corrosion production BT-FSW joint can be seen in Figure 3. Corrosion production of joints has the same composition, all of which consist of Al, C and O elements, which means that the main corrosion production of welded joints is $Al_2(CO_3)_3$ and Al_2O_3. Because the T6 6005A aluminum alloy contains two strengthening phases of Al-Fe-Si and Mg_2Si, and the corrosion production contains a small amount of Fe, Cl and Si element in addition to Cu_2O. As the potential of the matrix is about -700 mV, the Mg_2Si phase (-1200 mV) with lower corrosion potential is preferentially corroded as the anode end due to the action of the microcouple[8]. The corrosion potential of the Al-Fe-Si phase is higher (-200 mV) that the surrounding substrate will preferentially corrode to form pitting holes in the corrosion process, but it is not easy to corrode itself and trend to stay in the corrosion pit. Surrounding corrosion production is still in the state of "expansion-shedding". According to EDS spectrum analysis, after corrosion 120 h, the strengthening phase particles in the parent metal region are still retained in the corrosion pit to aggravate the corrosion, while the micro-galvanic effect is weakened with the shedding of welded joints second phase particles, which explains the reason why the macroscopic morphology of welded joints Nugget is less corrosive. At the same time, the pre-welded polished joint is similar to the composition of unpolished joints corrosion production, indicating that the corrosion resistance of the two groups is similar after the corrosion for 240h. It is consistent with the results of the weight loss rate of Figure 1.

3.2. Electrochemical corrosion pitting potential

In addition to chemical corrosion, the metal surface corrodes electrolyte solution and a small amount of carbon in the metal will form micro-battery, causing electrochemical corrosion. The electrochemical tests were carried out in different areas of the welded joints in corrosion soaking time with 30 min and 336 h. The corresponding polarization curve is shown in Figure 4. The electrochemical parameters are shown in Table 3. The self-corrosion current I_{corr} and the corrosion breakdown potential E_{corr} of polarization curve can be reflecting corrosion resistance. The larger the I_{corr} in the same corrosion system, the faster the electrode reaction and the corrosion rate. The higher the potential of E_{corr}, the more difficult loss of electrons of surface oxide film, the less the corrosion tendency[9]. Therefore, the polarization curve at the top left of the image shows stronger corrosion resistance of the material.

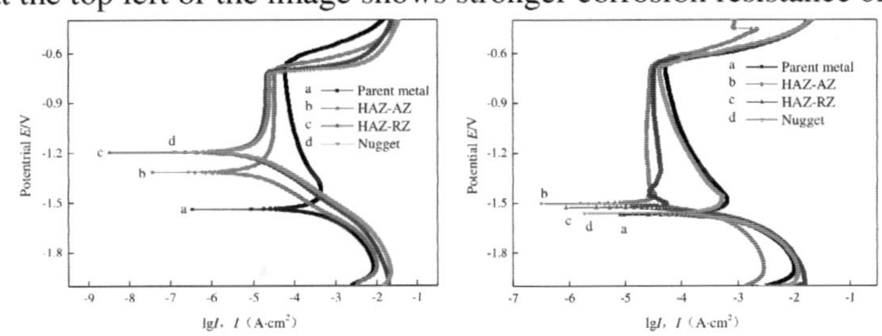

(a) Corrosion 30min pre-welded unpolished joints (b) Corrosion 30min pre-welded polished joints

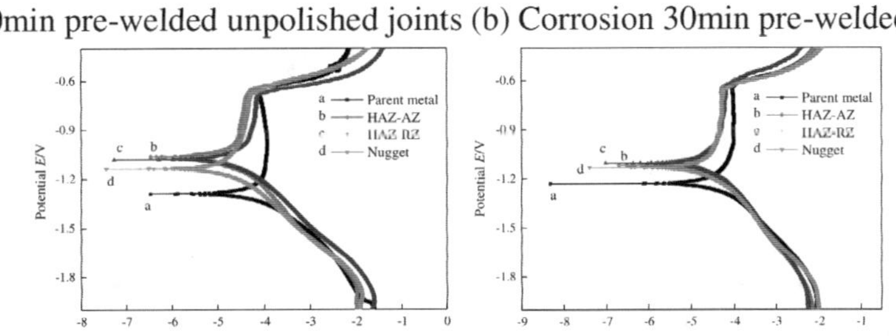

(c) Corrosion 336h pre-welded unpolished joints (d) Corrosion 336h pre-welded polished joints

Figure 4. Polarization curves of each zone of BT-FSW with different surface state

The polarization curve of the two kinds of joints has a large difference in Figure 4. When corrosion for 30 min, the polarization curve are located at the upper left of the whole image on both sides of the HAZ and weld nugget of unpolished joints, and the curve of parent metal is located at the lower right of the image. The electrochemical corrosion resistance of unpolished joints is higher than that in the parent metal region in the simulated acid rain environment. At this time, the corrosion potential of each zone of pre-welded polished is similar to that of parent metal, and polarization curve is located at the lower right corner of the image, indicating that the corrosion resistance of the polished joint is less than that of the unpolished joints. When the corrosion for 336 h, polarization curve in each area of unpolished joints has a small change compared with 30 min, while pre-welded polished joint polarization curve has a tendency to move to the left, which indicates that the corrosion resistance of the polished joint is weak when the corrosion time is short. However, as the corrosion time prolongs, the surface of the polished joints covered by the stable corrosion products layer. The thicker corrosion production are tightly bonded to the substrate, and the porosity of the film layer is low, so the corrosion resistance is improved to the extent similar to unpolished joints[10].

Combined with the data in Figure 4 and Table 3, it can be seen that the corrosion resistance of the BT-FSW joint is the best on both sides of HAZ, showing a lower self-corrosion density and corrosion breakdown potential. Followed by Nugget, its' self-current density and corrosion breakdown potential is similar to HAZ and exhibits a high corrosion sensitivity only when it is only corroded for 30 min. The parent metal is worst at the same corrosion time. When surface state is taken into consideration, the electrochemical parameters of unpolished joints are more stable than that of pre-welded, presumably because the pre-weld unpolished process retains the dense Al_2O_3 film on the surface. As an solid-state welding method, the Al_2O_3 film on the shoulder contact surface is broken into oxide particles under intense friction heat input, which participates in the plastic flow process of the joint material and finally remains in the joint structure. The oxide particles still have a certain protective effect, which improves the corrosion breakdown potential of the oxide film on the surface of the joint. In addition to the parent metal of well-preserved Al_2O_3 film, the corrosion potential of the unpolished joints should be higher than the pre-welded polished joint of the corresponding area, which is consistent with the distribution of the two surface state welded joints polarization curve when corroded for 30 min. When the corrosion exceeding 336 h, the thickness of corrosion production on the surface of the test piece increases, which reduces the adsorption of Cl^- SO_3^- and HCO_3^-on the surface of the body. so the corrosion potential of the pre-welded polished state joint is increased to the same level as the pre-weld unpolished joints, and the corrosion resistance is equivalent.

Table 3. Electrochemical parameters of BT-FSW joints with different surface state

Test sample number	Corrosion time	Current density/ potential	Parent metal	Heat affected advancing side	Heat affected retreating side	Nugget
Pre-weld unpolished joints	30 min	$I_{corr}/\mu A{\bullet}cm^{-2}$	529.16	15.11	9.54	9.01
		E_{corr}/mV	-1561.35	-1311.75	-1206.07	-1162.43
	336 h	$I_{corr}/\mu A{\bullet}cm^{-2}$	46.74	13.47	20.71	13.50
		E_{corr}/mV	-1288.89	-1053.97	-1060.22	-1124.05
Pre-welded polished joints	30 min	$I_{corr}/\mu A{\bullet}cm^{-2}$	531.33	33.62	111.17	492.31
		E_{corr}/mV	-1565.51	-1491.29	-1525.36	-1556.76
	336 h	$I_{corr}/\mu A{\bullet}cm^{-2}$	29.75	16.46	14.93	16.07
		E_{corr}/mV	-1227.92	-1115.61	-1131.94	-1097.41

3.3. Electrochemical AC impedance analysis

The AC Impedance Test (EIS) is a test method for the corresponding corrosion mechanism by measuring the response signal generated by the small peak AC signal perturbation electrochemical system. Figure 5 shows the Nquist map and Bode map in open circuit potential (-0.78 V) under the simulated acid rain environment of welded joints. It can be seen from Figure 5(a) that the fitting curve and the test curve have higher coincidence in the frequency test range. The Nquist spectrum has consistent characteristics, showing two incomplete semicircular capacitive arcs. The high-frequency capacitive anti-arc on the left side represents the nature of the self-passivation film's own impedance, and the low-frequency capacitive anti-arc on the right side represents the nature of the impedance of the workpiece electrode after the passivation film is dissolved when the corrosion diffusion channel is completely opened[11]. The impedance of the passivation film is related to the arc. The larger the radius of the arc, the greater the reaction resistance, and the less the charge is transferred[12]. Judging from the radius of the arc, the low-frequency capacitive reactance of parent metal is much larger than that of the two surface state welded joints. The capacitance resistance of unpolished joints is slightly larger than that of pre-welded polished. The value ($|Z|_{0.01Hz}$) of parent metal in low frequency impedance mode is between 10^5~10^6 $\Omega{\cdot}cm^2$, which is higher than that of the two welded joints. It is presumed that a new film layer formed on the surface of Nugget after the BT-FSW is completed. The new film layer is thinner than the parent metal passivation film, and the corrosion breakdown potential of this film is lower than that of base metal passivation film. The thicker parent metal passivation film acts as a shielding layer at the initial stage of corrosion, which well isolates the contact between the corrosive medium and the substrate, and plays a good protective role. The results show that the corrosion resistance of pre-welded polished joint was the worst at the beginning of corrosion, which is consistent with the results of salt spray test and Tafel curves.

The frequency-impedance of parent metal and joints is approximately a straight line with slope -1 in Figure 5(b), indicating that the resistance and capacitance are in parallel. Figure 5(c) shows that the Bode spectrum has two peaks, standing two time constants. So the EIS spectrum is fitted by the equivalent circuit diagram shown in Figure 6, and the fitting parameters of circuit component are obtained as shown in Table 4. R_s represents the resistance of the electrolyte solution between the test piece and the calomel electrode; R_f represents the impedance of the passivation film; R_{ct} represents the charge transfer resistance between the contact interface between the passivation film and the electrolyte; Q_c is common phase angle component; C represents the interface capacitance between the passivation film and the electrolyte interface[13]; n is the degree of similarity with the phase angle component Q_c and the ideal capacitance; *CHI* is the total variance of the fitted curve.

It can be seen from Table 4 that the equivalent circuit diagram fitting curve error does not exceed 7×10^{-3} and reflect the mechanism of the corrosion process. Since the EIS impedance spectrum is measured in the acid solution of the same corrosive medium, the value of electrolyte solution resistance R_s is less different. The R_f value and R_{ct} value of parent metal are significantly higher than the two surface state welded joints. The impedance of the parent metal passivation film is large when the corrosion reaction occurs, and so as the charge transfer resistance of the electrolyte and the workpiece surface, indicating the corrosion resistance of parent metal is the best. The corrosion resistance of the specimen: parent metal>unpolished joints>polishing joint is from strong to weak when the corrosion medium is exposed for a short time.The EIS fitting results show that the polishing process will reduce the corrosion resistance of the joints. But when the corrosion time is prolonged, the surface of the joints will form a protective layer of stable corrosion production, which hinders the contact channel of corrosive ions, The corrosion resistance of the joints is not obviously reduced by grinding process.

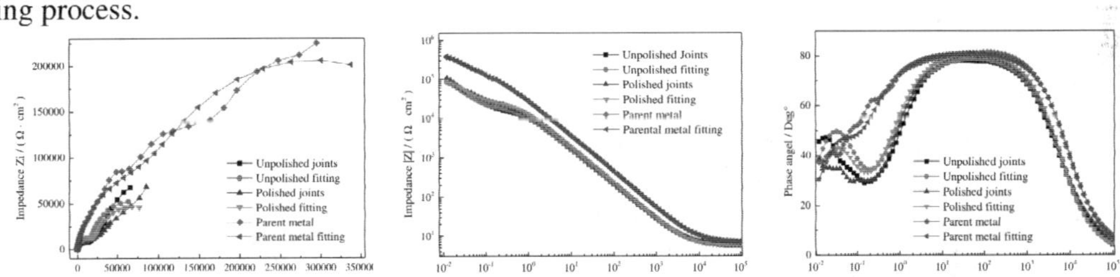

(a) Nquist map; (b) Bode map (impedance-frequency); (c) Bode map (phase angle-frequency)

Figure 5. AC impedance spectra of welded joints in different surface state

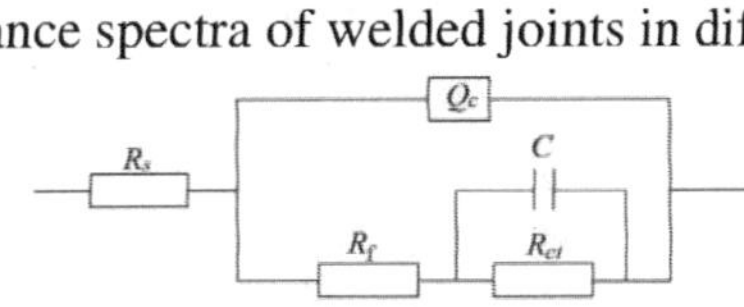

Figure 6. Equivalent circuit diagram of impedance spectrum of welded Joint

Table 4. EIS fitting data of welded joints in different surface state

Test sample	R_s ($\Omega\cdot cm^2$)	Q_c ($\Omega^{-}\cdot cm^{-2}\cdot s^n$)	n	R_f ($\Omega\cdot cm^2$)	C (μF)	R_{ct} ($\Omega\cdot cm^2$)	*CHI*
Unpolished joints	5.609	1.513×10^{-5}	0.894	2.178×10^{4}	122.4	1.02×10^{5}	6.59×10^{-3}
Polished joints	5.508	1.216×10^{-5}	0.911	2.906×10^{4}	109.2	0.87×10^{5}	6.19×10^{-3}
Parent metal	6.951	0.655×10^{-5}	0.907	1.851×10^{5}	22.59	3.26×10^{5}	2.16×10^{-3}

IOP Conf. Series: Materials Science and Engineering **727** (2020) 012003 doi:10.1088/1757-899X/727/1/012003

4. Conclusion

(1) The rate change of the salt spray test accords with the law of "lower-rise-lower". Before 120 h, the weight loss rate of parent metal is the minimum, the rate of pre-welded polished joints is the highest and the most corroded. After 240 h, the corrosion rate of parent metal rises rapidly. At this time, the weight loss rate of the two welded joints is similar, and better than that of parent metal.

(2) The surface of salt spray test joint has shallow pitting corrosion, mainly shallow exfoliation. The active anion in acidic medium and the high-potential second phase particles not falling off in the corrosion pit are the main corrosion driving forces of pitting corrosion.

(3) Comparing the different dynamic potential polarization curves: short-term corrosion for 30 min, the self-corrosion current of unpolished joints is lower, the corrosion breakdown potential of the passivation film is higher, the corrosion resistance is much higher than that of parent metal. The pre-welded polished joints show poor corrosion resistance. After corroded for 336 h, the pre-welded polished has limited influence on electrochemical corrosion resistance

(4) The electrochemical characteristics of the EIS spectra of the two kinds of joints are basically the same, and the Nquist spectra reflect the corrosion behavior of the passivation film on the surface of the two joints is also consistent. At short time corrosion, the oxide film's resistance and charge transfer resistance on the surface of parent metal are much higher than the new film formed on the surface of the two welded joints. Meanwhile, the pre-welded polished process reduces the corrosion resistance of the joint to some extent. As the corrosion time prolongs, the pre-weld surface state has little effect on the corrosion resistance of the joint. Therefore, the process of production can be considered to cancel the pre-welded polished process to reduce the production cost.

References

[1] Sheng X F, Yang W C.(2012) Effect of artificial ageing on intergranular corrosion of 6005A al alloy. J. *The Chinese Journal of Nonferrous Metals*. 22(08): 2174-2180.

[2] Wu H X, Yang L.(2014) Development status of aluminum profiles applied on China's rail vehicle. J. *Light Alloy Fabrication Technology*. 42(01): 18-20.

[3] Shen H, Dong C F. (2013) Effect of pH on electrochemical behavior of 2024-T351 aluminum alloy in NaCl solution. J. *Corrosion & Protection*. 34(02): 107-110.

[4] Li T. (2011) Effect of Cl^- concentration on electrochemical behavior of 2A12 aluminum alloy. J. *Corrosion & Protection*. 32(04):256-260.

[5] Li J F. (2005) Function machanism of secondary phase in local corrosion of aluminum alloy. J. *Mater Rep*. 19(2):81-83.

[6] Hu B H, Xing L. (2012) Corrosion morphology of friction stir weld of LY12 aluminum alloy in 3.5%NaCl solution. J. *Journal of Nanchang Institute of Aeronautical Technology*. 26(03):52-56.

[7] Harlow DG, Wei RP. (1998) A probability model for the growth of corrosion pits in aluminum alloys induced by constituent particles. J. *Engineering Fracture Mechanics*. 59(3):305-325.

[8] Zhang J, Zhang T. (2014) Crevice corrosion behavior of 5083 and 6061 aluminum alloys. J. *Corrosion Science and Protection Technology*. 26(02):125-131.

[9] GHALE. (2010) Corrosion rsistace of aluminum and magnesium alloy. In: John, W.S. (Eds), *Undersanding, performance, and texting*.New jersey. pp. 180.

[10] Liu J, Zhang P. (2013) Corrosion behavior of anodized film of 7075 aluminum alloy. J. *Mater Protec*. 46(08):56-57.

[11] Liu J, H, Li Y X, Yu M et al. (2011) Effects of chromium-free sealing methods on corrosion resistance of anodic oxide films on aluminum alloy. J. *T Mater Heat Treat*. 32(10):129-133.

[12] XU J, HUANG H J, LI Z Y, et al. (2016) Corrosion behavior of a ZrCN coated Ti alloy with potential application as a bipolar plate for proton exchange membrane fuel cell. J. *Journal of Alloys and Compounds*. 663: 718-730.

[13] MANI P S, RAJENDRAN N. (2017) Corrosion and interfacial contact resistance behavior of electrochemically nitrided 316L SS bipolar plates for proton exchange membrane fuel cells. J. *Energy*. 133: 1050-1062.

IOP Conf. Series: Materials Science and Engineering **727** (2020) 012004 doi:10.1088/1757-899X/727/1/012004

Preparation and Properties of UV Irradiation Modified ZnO-HA Coating on the Biomedical Magnesium Alloys

Jin'e Sun[1,*], Jie Liu[2], Zhanxiang Niu[1], Meizheng Chi[1], Xuecheng Peng[1], Yong Yang[3] and Qihui Lin[4]

[1] Tianjin College, Beijing University of Science and Technology, Tianjin 301800, People's Republic of China

[2] Fourth Municipal Administration Co. LTD, Tianjin 300000, People's Republic of China

[3] Equipment Bidding Center, Naval Armament Department, Beijing 100071, People's Republic of China

[4] Beijing Electrical Engineering School, Beijing 100051, People's Republic of China

[*]Corresponding author's Email: sunjine0406@163.com

Abstract. In order to improve the corrosion resistance of magnesium alloys, ZnO-HA coating was prepared on the surface of magnesium alloys by using ultraviolet (UV) irradiation and hydrothermal method. The effects of ultraviolet irradiation time on the microstructure, the electrochemical properties and corrosion resistance of ZnO-HA coating were studied. The results indicated that the ZnO-HA coating obtained by 24 h of UV irradiation combined hydrothermal method was dense and uniform, without any cracks, and it was mainly composed of nanoparticle-like HA and flake-like ZnO. Electrochemical properties and corrosion resistance of ZnO-HA coated magnesium alloys were studied in simulated body fluid (SBF). Compared with the bare magnesium alloys, the coated magnesium alloys have good corrosion resistance. In vitro immersion experiments, ZnO-HA coating could provide good protection for the magnesium alloys substrate, indicating that ZnO-HA coating prepared by UV irradiation combined hydrothermal method can be used as a potential modified biomaterial for magnesium alloys.

1. Introduction

As the degradable biomedical materials, magnesium alloys have been praised as the third generation of biomedical materials in the world due to their elastic modulus similar to natural bone, excellent biodegradability, good strength and low toxicity [1]. However, too rapid degradable rate of magnesium alloys in complex physiological environment [2] leads to the premature failure of mechanical properties and the excessive bone healing time, which has become the major bottleneck limiting their clinical application. Therefore, it is very important to improve the corrosion resistance of magnesium alloys to prolong their service life in the human body.

Surface modification with bioactive coating is one of the effective methods to reduce degradation rate and enhance the bioactivity of magnesium alloys. Hydroxyapatite (HA) [3], as a biodegradable modified material, is quite similar to the mineral composition and structure of natural bone and plays an important role in improving the corrosion resistance of magnesium alloys, promoting cell adhesion, proliferation and differentiation, and new bone formation. Zinc oxide (ZnO) is a semiconductor oxide, which can stimulate a large number of active hydroxide radicals (·OH) on the surface of ZnO by UV irradiation, and improve the hydrophilicity and the biological activity of the material [4]. In addition, ZnO has almost same band gap energy (3.2 eV) as titanium dioxide (TiO_2), and its photocatalytic capacity is similar to that of TiO_2. In the latest research, Liu et al. [5] stimulated nano TiO_2 coating to generate hydroxide radicals (·OH) on its surface by UV irradiation, which improved its hydrophilicity and further enhance the quick formation of the bioactive apatite layer. Moreover, the *in vivo* results indicated that there were mineralized apatite and bone tissue formed around the implants after implanting into the animals for 2 months. Compared with TiO_2, ZnO is considered to have higher photocatalytic efficiency due to its higher quantum efficiency [6]. Therefore, more hydroxide radicals (·OH) could be generated on the surface of ZnO by using UV irradiation, and promote the heterogeneous nucleation and growth of HA [7]. As a consequent, it is reasonably believed that ZnO-HA coating on magnesium alloys surface modified by UV irradiation is expected to improve the corrosion resistance of magnesium alloy, promote the formation of new bone, significantly shorten the time of bone integration, and finally obtain the optimal implantation effect.

In order to make the coating possess favorable protection effect for magnesium alloys, in this present work, ZnO-HA coating was successfully prepared on the magnesium alloys surface by the UV irradiation modified technology combined hydrothermal method, and the microstructure, the electrochemical properties, corrosion resistance, and in vitro degradation behavior of ZnO-HA coating on the magnesium alloys were investigated.

2. Materials and experimental methods

2.1 Preparation of ZnO-HA coating on magnesium alloys

0.01782 g $Zn(NO_3)_2 \cdot 6H_2O$ was mixed with 20 ml 28 % ammonia water to prepare the mixture solution. Measuring 37.5 ml of the mixture solution to the reaction kettle, in which the treated AZ31 magnesium alloys samples were placed. Afterwards, ZnO coating was prepared on magnesium alloys surface by hydrothermal method at 95 °C for 10 h. The resultant samples were divided into three groups and irradiated by UV light for 0 h, 12 h and 24 h, respectively and the intensity of UV irradiation was 5 uW/cm^2. After that, they were put into the reaction kettle with 37.5 ml of the mixed solution, in which EDTA2NaCa was as the calcium source and Na_2HPO_4 was as the phosphorus source and the concentrations of Ca^{2+} and PO_4^{3-} in the mixed solution were both 0.20 mol/L. ZnO-HA coating was prepared on the magnesium alloys surface by hydrothermal method at 120 °C for 13 h, finally, they were defined as UV0h-ZnO-HA, UV12h-ZnO-HA and UV24h-ZnO-HA sample, respectively.

2.2 Coating characterization and tests

The phase and surface morphologies of the obtained coatings were characterized using X-ray diffraction (XRD, D/Max-2500 Rigaku, Japan) with Cu Kαradiation in the range of 5-70 °(2θ values) and scanning electron microscopy (SEM, S-4800, Hitachi, Japan). To evaluate the electrochemical property of samples, the coated magnesium alloys were examined by potentiodynamic polarization test in SBF solution at 37 °C based on a CHI660C electrochemical workstation with a typical three-electrode cell. In this test, 3 samples were investigated to take the average. Moreover, the prepared ZnO-HA coated magnesium alloy samples were soaked in simulated body fluids (SBF) at 37 °C for different periods for assessing the corrosion behavior *in vitro*. The pH value of SBF solution was measured and recorded on time every day (24 h), and the SBF solution was changed every 2 days.

3. Results and discussion

3.1 Composition and microstructure of coatings

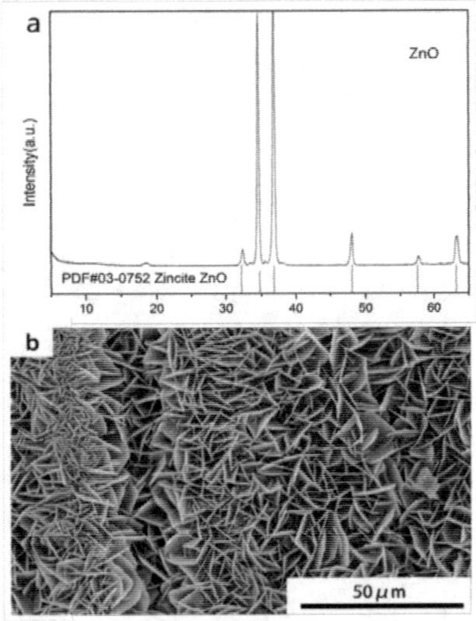

Figure 1. XRD pattern (a) and SEM morphology (b) of ZnO coated magnesium alloys

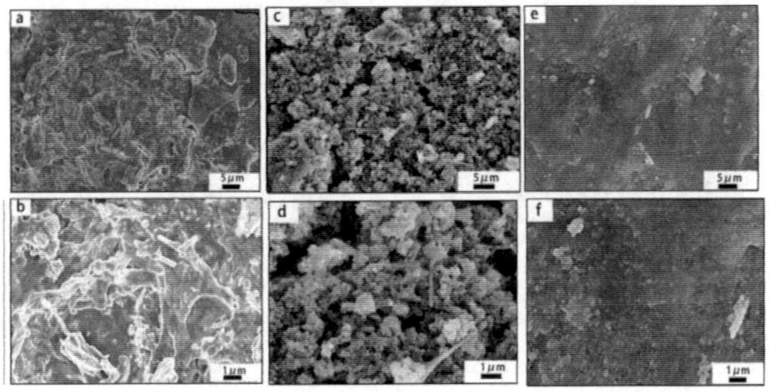

Figure 2. SEM morphologies of coated samples with different UV-irradiated time (a, b: 0 h; c, d: 12 h; e, f: 24 h) after the hydrothermal method for 13 h.

Figure 1. showed the XRD pattern and SEM morphology of ZnO coated magnesium alloys samples. Obviously, the diffraction peaks detected were consistent with the diffraction peaks of standard card ZnO (JCPDS. No 03-0752) (figure 1. (a)). As shown in figure 1. (b), ZnO coating was composed of irregular flakes growing on the magnesium alloys substrate in a vertical and random direction, these flakes distributed relatively uniformly and their size ranged from about 200 nm to 2 μm, which formed the micro-nano structural ZnO coating. In addition, some small gaps existed among the irregular plates, constructing a porous network of micro-nano structure, which could effectively improve the hydrophilicity of the coating surface. Zhang et al. [8] has indicated that under UV irradiation, compared with acicular ZnO, micro-nano flake-like ZnO could produce more active hydroxyl groups (·OH) and have stronger hydrophilicity, providing a good foundation for subsequent preparation of HA coating on the magnesium alloy surface.

Figure 2. presented SEM morphologies of coated samples with different UV-irradiated time of 0 h, 12 h and 24 h after the hydrothermal method at 120 °C for 13 h. It could be observed that the surface morphologies of coated samples displayed an enhanced densification with the extension of UV irradiation time. In the absence of UV irradiation, the surface of the coated sample was uneven and irregular. After 12 h of UV irradiation, the coating on the sample was consisted of spherical shape particles, and there were some pores with uneven size around them, ranging from 1 to 5 μm. Continuing to increase UV irradiation time to 24 h, the coating surface of sample was mainly composed of uniform small particles with the size of about 100 nm. Moreover, the coating texture became more uniform and flat without any obvious pores in comparison with those after UV irradiation of 0 h and 12 h. In addition, the XRD pattern of the UV24h-ZnO-HA sample was shown in figure 3. It could be evidently found that the diffraction peaks in the XRD pattern were consistent with the standard card HA (JCPDS No. 86-0740). Besides, the diffraction peaks of ZnO (JCPDS No. 03-0752) were also detected, indicating that ZnO-HA coating was successfully prepared on the surface of magnesium alloys by the UV irradiation of 24 h combined hydrothermal method, and the diffraction peaks and their intensity revealed that the prepared HA had good crystallinity.

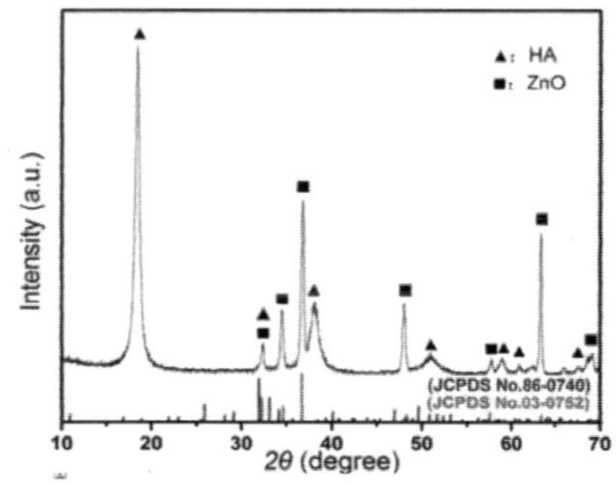

Figure 3. XRD pattern of coated sample with UV irradiation of 24 h after the hydrothermal method for 13 h.

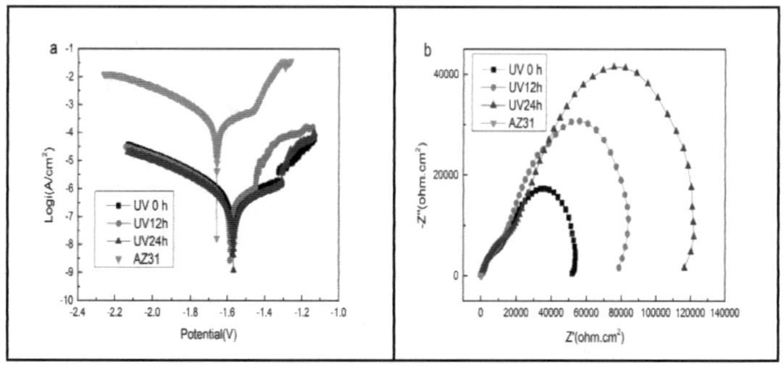

Figure 4. Electrochemical properties of ZnO-HA coated and naked samples in SBF solution (a: Potentiodynamic polarization curves; b: Nyquist plots)

3.2 Corrosion resistance

Table 1. Corrosion properties of ZnO-HA coated and naked samples

Samples	E_{corr} (V)	i_{corr} (I_C/A·cm^{-2})	R_t (Ω·cm^2)
AZ31	-1.6658	56.400 × 10^{-7}	0.05 × 10^4
UV0h	-1.5913	2.764 × 10^{-7}	5.00 × 10^4
UV12h	-1.5824	1.994 × 10^{-7}	7.80 × 10^4
UV24h	-1.5736	0.973 × 10^{-7}	11.50 × 10^4

In order to evaluate the influence of UV-irradiated time on corrosion properties of ZnO-HA coating, the electrochemical measurements of ZnO-HA coated and naked magnesium alloys in SBF were performed and the results were revealed in figure 4 and Table 1. With the prolonging of UV irradiation time, both of the corrosion potential and charge transfer resistances of coated samples gradually increased, and the corrosion current density gradually decreased. Among them, the E_{corr} of UV24h-ZnO-HA samples was the largest (-1.5736 V), and its i_{corr} was the smallest (0.9737 × 10^{-7} A·cm^{-2}), which reduced by an order of magnitude compared with those of UV0h-ZnO-HA and UV12h-ZnO-HA samples. Furthermore, its R_t reached 11.50 × 10^4 Ω·cm^2 that was in exceed of those of UV0h-ZnO-HA and UV12h-ZnO-HA samples, even improved the three orders of magnitude with respect to naked magnesium alloys. Generally speaking, the greater the corrosion potential is, the smaller the corrosion current density is, and the greater the charge transfer resistance is, implying that the test sample has favorable anti-corrosion property [9]. Therefore, it could be concluded that UV24h-ZnO-HA sample had good corrosion resistance. According to the analysis from figure 2, the micro-cracks and pores on the surface of UV0h-ZnO-HA and UV12h-ZnO-HA samples were the main reasons leading to the decline of electrochemical performance. Nevertheless, as for UV24h-ZnO-HA sample, ZnO-HA coating with high density, uniformity and integrity could be used as a good physical barrier to prevent the penetration of aggressive ions in SBF into magnesium alloys and improve the corrosion resistance of magnesium alloys [8, 10]. Consequently, UV24h-ZnO-HA sample was selected as the representative samples to further study the degradation behavior in SBF for different time.

3.3 Degradation behavior in vitro

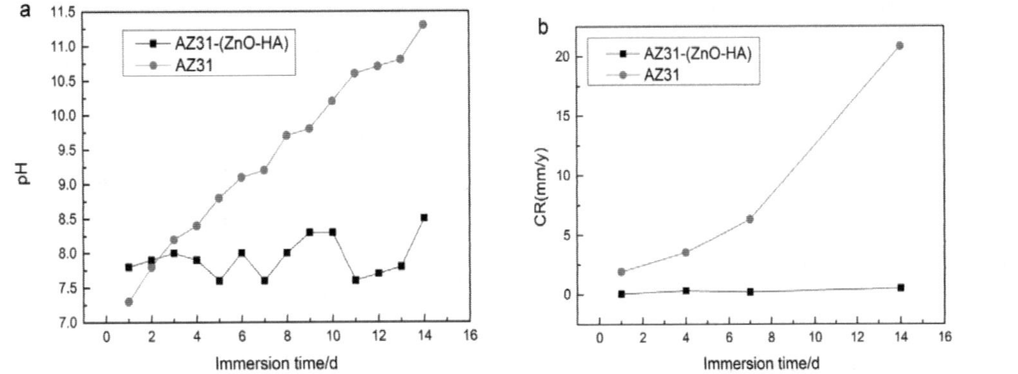

Figure 5. (a) pH variation ; (b) corrosion rate of ZnO-HA coated samples with UV irradiation for 24 h after soaking in SBF solution for different times

The pH variation of naked magnesium alloys and ZnO-HA coated samples immersed in SBF solution for different time were illustrated in figure 5. (a). During the immersion of 14 days, the pH value of naked magnesium alloys in SBF rose linearly, and exceeded 11 on the 14th days. On the contrary, pH value of coated samples fluctuated a little with the increasing of immersion time, and the overall pH value was relatively stable, ranging from 7.5 to 8.5, which was primarily owing to the synergistic effect between HA and Ca^{2+} and OH^- ions, that was, ions exchange and neutralization reaction would occur when Ca^{2+} and OH^- ions were adsorbed by HA, so that OH^- was also consumed when PO_4^{3-} on the protonized surface reacted with Ca^{2+}, as a result, new mineralized layer was gradually formed on the surface of coated samples, thereby inhibiting the pH increase in solution environment [11]. Meanwhile, the corrosion rate of coated sample and naked sample soaked in SBF for 1, 4, 7 and 14 days was further analyzed as shown in figure 5. (b). During the whole soaking process, the corrosion rate of naked magnesium alloys was obviously higher than that of coated samples and its corrosion rate was 20.8 mm/y when immersed to 14 days. However, the corrosion rate of the coated samples was more stable, only 0.48 mm/y on the 14th day of immersion, which met the requirement that the corrosion rate of the biodegradable materials in SBF solution should be less than 0.5 mm/y [6]. ZnO-HA coated samples exhibited good corrosion resistance during 14 days of immersion, further illustrating that ZnO-HA coating played an important role in protection for magnesium alloys during the whole process of soaking in SBF, specifically, ZnO-HA bilayer as two solid barrier, effectively blocked the erosion of erosion ions in SBF solution. Besides, the good corrosion resistance might be linked to the formed mineralizated products on the coating surface. Relative studies have also demonstrated that the new mineralizated products formed on the coating surface due to the presence of HPO_4^{2-}, Ca^{2+}, and PO_4^{3-} in SBF made up for cracks on the surface of coated samples and hindered the corrosion of magnesium alloys [8]. As a consequence, it could be inferred that the ZnO-HA bilayer on the surface of the coated sample and the rapidly formed mineralized layer during the soaking process effectively protected the magnesium alloys substrate together, improved the corrosion resistance of the magnesium alloys, and slowed down the corrosion rate of ZnO-HA coated samples.

4. Conclusion

In this paper, a compact ZnO-HA coating was prepared on the surface of magnesium alloys by UV irradiation combined hydrothermal method. When the UV irradiation time was 24 h, the coating surface was dense and uniform. Electrochemical measurement results indicated that the coated magnesium alloys with UV irradiation of 24 h had good corrosion resistance, and the corrosion potential was the greatest (1.5736 V), the corrosion current density was the smallest (0.9737×10^{-7} $A{\cdot}cm^{-2}$), the charge transfer resistance was the greatest (11.50×10^4 $\Omega{\cdot}cm^2$). In addition, during 14 days of *in vitro* immersion test, the pH value of coated samples in SBF solution remained within the range of 7.5-8.5, and the corrosion rate of the coated samples was only 0.480 mm/y, indicating that the ZnO-HA coated samples had a good protective effect on the magnesium alloy substrate, effectively improved the corrosion resistance of magnesium alloys, and slowed down the degradation rate. Therefore, UV radiation combined hydrothermal method could be acted as a potential technology to improve corrosion resistance of biodegradable magnesium alloys.

Acknowledgements
The work was financially supported by the Scientific Research Project of Tianjin Education Commission (Grant No. 2018KJ264)

References
[1] Wang G, Li J and Zhang W 2014 *Int. J. Nanomed* **9** 2387–98
[2] Wang M J, Chao S C and Yen S K 2016 Corros. Sci **104** 47–60.
[3] Shen S B, Cai S and Ling R 2017 Chem. Eng. J. **309** 278–87
[4] Feng W, Jing G L and Zhang G P 2008 *Chemosphere* **72** 407
[5] Liu X Y, Zhao X B, Li B and Cao C 2008 *Acta. Biomater* **3** 544–52
[6] Chen X, Shen S, Guo L and Mao S 2010 *Chem. Rev* **11** 6503–70
[7] Shen S B, Cai S, Li Y and Ling R 2017 *J. Chem. Engi* **309** 278–87
[8] Zhang Y 2018 *J. Mater. Sci. Mater. Electron* **29** 1576–83
[9] Amaravathy S, Sathyanara S and Sowndarya N 2014 *Ceram. Int* **40** 6617–30
[10] Hornberger H, Virtanen S and Boccaccini A R 2012 *Acta. Biomater* **8** 2442–55
[11] Chen Y, Xu Z and Smith C 2014 *Acta. Biomater* **10** 4561–73

IOP Conf. Series: Materials Science and Engineering **727** (2020) 012005 doi:10.1088/1757-899X/727/1/012005

Using KH590-Polybutadiene as the New Compatilizer to Improve Carbon Black / Ethylene-Propylene-Diene Terpolymer Composites

Weili Wu* **and Bowen Yu**

College of Materials Science and Engineering, Qiqihar University, Qiqihar, 161006, China

*wuweili2001@163.com

Abstract. Rubber materials could be widely used with the reinforcement of inorganic materials, the combination of inorganic materials and rubber needs compatilizer that play an important role in the polymeric composites. The ethylene-propylene-diene terpolymer (EPDM) has considerable radiation stability for gamma or neutron radiation, so it can be applied to cable insulating materials. The EPDM has to be reinforced with carbon black (CB) of inorganic fillers, which are hardly compatible with the EPDM matrix due to EPDM saturated bonds. In this work, The EPDM matrix modified with (3-Mercaptopropyl) trimethoxysilane (KH590)-polybutadiene (KHBR) was studied in details, and the changes in the mechanical, thermal and rheological properties and curing kinetics were investigated by tensile tests, dynamic thermal mechanical analysis (DMA), thermogravimetric analysis (TG), differential scanning calorimetry (DSC) and moving die rheometer (MDR). The results showed that the addition of KHBR improved the mechanical and thermal properties of the EPDM, and the optimized content was 2.5%.

1. Introduction

Rubber materials could be widely applied with the reinforcement of inorganic materials or inorganic fillers [1]. EPDM is a polymeric material, of which is usually filled with many kinds of organic or inorganic materials to improve the comprehensive performance for various application purposes [2,3]. EPDM has considerable radiation stability for gamma or neutron radiation, so it can be applied to cable insulating materials [4,5]. However, EPDM has to be reinforced with carbon black (CB) of inorganic fillers, which are hardly compatible with the EPDM matrix due to EPDM saturated bonds [6]. Therefore, the use of coupling agents is inevitable for the blend of inorganic fillers and EPDM [7,8].

The CB is easily to adsorb polybutadiene because of its unsaturated bonds[9,10]. As a result, in this work, KH590-polybutadiene (KHBR) was used as a compatilizer to improve interaction between EPDM and carbon black.

2. Experimental

2.1. Materials

EPDM as grade 23080P and polybutadiene (BR) as grade BR-9000 were produced by Chinese Jilin chemical industrial inc.. Dicumyl peroxide (DCP) as chemical pure was supplied by Shanghai chemical reagents company of Chinese medicine group. The other ingredients, such as stearic acid (CA), 2-mercaptobenzothiazole (accelerant M), grease, sulfur (S), (3-Mercaptopropyl) trimethoxysilane (KH590), carbon black (CB) and so on, are all common commercially avail materials obtained in the market.

2.2. Sample preparation

First, the BR was masticated on SK-160 two-roll mill (electrical machinery factory of Tianjin) at 45 °C and the nip gap about 1.0 mm for 10 min, and hydrolysis KH590 was added into the BR, in which the mass ratio of KH590 and BR is 1:10, then mixed with the S and half of the CB to product HK590-modified BR (KHBR).

Second, the KHBR was mixed into EPDM matrix, and the mixed rubber was prepared by adding various ingredients, the order of process was EPDM (raw rubbers masticated) → ZnO, CA, accelerant M, KHBR(1, 2.5 and 5 weight percent of EPDM content) →half of the CB → DCP, the preparation process of sample was showed in the Figure 1.

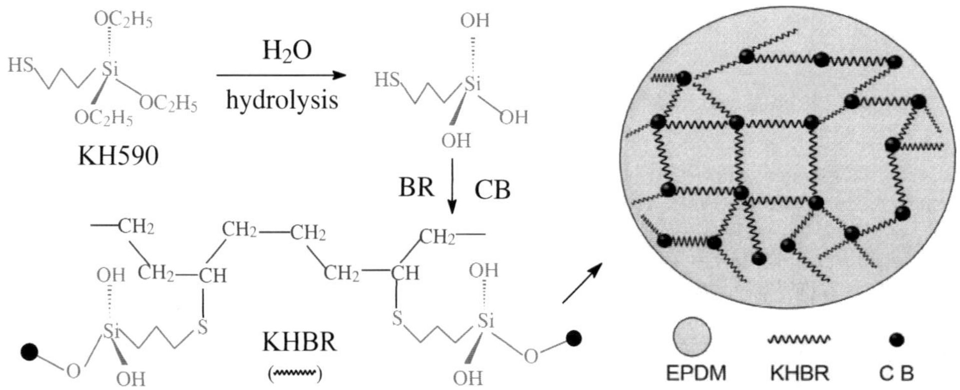

Figure 1. Reacting mechanism of CB / EPDM composites modified by KHBR.

The formulation of the mixed-EPDM rubber was given in Table 1. The compounds were mixed in a laboratory size (SK-160) mixing mill at a friction ratio of 1:1.1, carefully controlled temperature in the range of 50-60 °C, the nip gap was 2 mm for 20-30 min and uniform cutting operation. Then the mixture with different compositions was molded in an electrically heated hydraulic press (XLB-D350 ×350) at 180 °C for 20 (or t_{90} +15) min under 10MPa (optimum curing condition).

Table 1. Formulation of mixed-EPDM rubber (phr).

EPDM	CB	ZnO	Grease	CA	Accelerant M	DCP	S	KHBR
100	40	5	3.5	1.5	2	2	2	Variable

2.3. Characterization

Tensile tests were performed on dumbbell-shape specimens according to ISO 37-1994.

Shore A hardness was measured on the thickness of 6 mm according to ISO 48-1994 and the tear test was completed according to ISO 34-1994.

The samples were tested and the average of the values was taken. Seven samples were used for each measurement and an average of five readings was taken as the resultant value and removed the maximum and minimum values.

DSC thermal analysis was carried out with differential scanning calorimetry tester (Malvern 204F1, UK), nitrogen purging, and gas flow is set as 50 mL min^{-1}, the temperature ranged from -90 °C to -20 °C for the samples at a heat rate of 10 °C min^{-1}.

Dynamic thermal mechanical analysis (DMA) was performed using a DMA Q800 dynamic mechanical analyzer (TA instrument Co., USA) at a frequency of 1Hz in the temperature range from -100 °C to 0 °C with a heat rate of 5 °C min^{-1}. The storage modulus and tan δ parameters were obtained from DMA analysis. In addition, the thermal stability was analyzed from 15 °C and 550 °C with a TGA/STDA851 thermogravimetric analyzer (Mettler-Toledo Co., Swiss).

The rheological cure properties were determined with MZ 4010 B mark Moving Die Rheometer (MDR, Jiangsu Mingzhu Testing Machinery Co., Ltd. China).

3. Results and Discussion

3.1. Mechanical properties of KHBR-modified EPDM composites

The effects of KHBR content on the mechanical properties of EPDM composites are given as shown in Table 2. Tensile strength was increased with adding KHBR. It is also seen that elastic modulus and Shore A were improved with adding KHBR, the elongation at break decreased a little. The reasons can be interpreted that the addition of KHBR increased the compatibility between EPDM rubber and inorganic filler, resulting in the tensile strength, elastic modulus and Shore A hardness increased. However, when the addition of KHPB was above 5%, the combination of EPDM and CB was firm, increasing the rigidity of materials, so the elongation was falling.

Table 2. Mechanical properties of EPDM composite with adding KHBR.

KHBR (%)	Tensile strength (MPa)	Elastic modulus(MPa)	Elongation at break (%)	Shore A hardness
0	4.2	0.91	310	38
1	5.6	1.00	290	46
2.5	5.9	1.10	290	46
5	5.4	1.13	260	45

3.2. Rheological cure analysis

Curing kinetic and rheometer results are given in Table 3. M_H-M_L increased with the addition of the KHBR up to 2.5%. When the increase of the KHBR content get to 5%, the t_{90} was the highest, and the scorch time (t_{s2}) was minimum for the 2.5% KHBR content. The curing kinetic parameters are calculated and the results are given in Table 3.

Induction period (t_d) values calculated are given in Table 3. The decrease of the t_{fir} (time required for crosslinking to become unperturbed first-order reaction) showed that the crosslinking reaction started earlier with adding KHBR, the t_{fir} was the lowest for the 2.5% KHBR, confirming that the lowest scorch time with 2.5 % KHBR, M_H-M_L result with 2.5 % KHBR is the best, it suggests that the degree of curing is the best.

Table 3. Curing kinetic parameters.

	0% KHBR	1% KHBR	2.5% KHBR	5% KHBR
M_H-M_L (N.m)	0.4	0.4	0.5	0.4
t_{90} (sec)	260	260	260	270
t_{s1} (sec)	80	76	65	76
t_{s2} (sec)	120	114	83	121
t_d(sec)	79	76	70	81
t_{fir} (sec)	140	94	73	83

3.3. TGA of KHBR-modified EPDM composites

As is well known, DTG curve represents the weight loss rate of speed, the peak is the fastest temperature of weightloss and it can distinguish weightloss stage usually. DTG curves showed that the rate of weight loss was decreased with the addition of KHBR as shown in Figure 2, and the peak height decreased with the increase of KHBR content. It suggests that high or low KHBR content is bad for the thermal properties of EPDM composites. And the composites with 2.5% KHBR have better thermal stability.

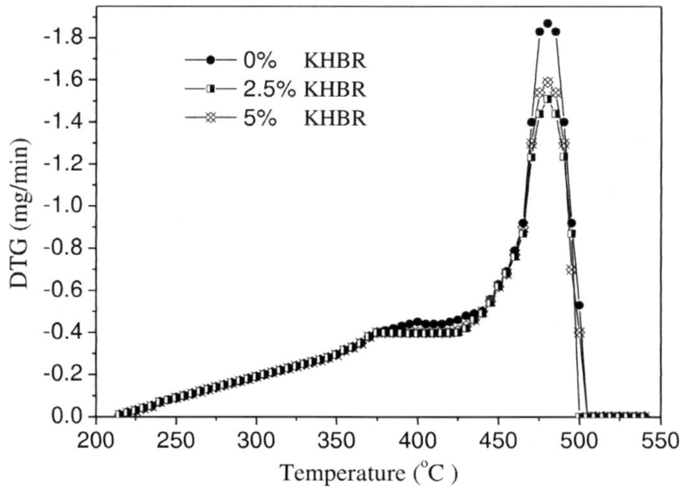

Figure 2. DTG of KHBR-modified EPDM composites.

3.4. DSC of KHBR-modified EPDM composites

DSC thermograms showed that the Tg of 0% KHBR was -56 °C and that of 2.5% KHBR is -61 °C as shown in Figure 3. It was clear that the addition of KHBR decreased the glass transition temperature. It could be considered that the decrease of glass transition temperature (Tg) was most probably due to addition of polybutadiene-based KHBR, which had a much lower Tg than EPDM.

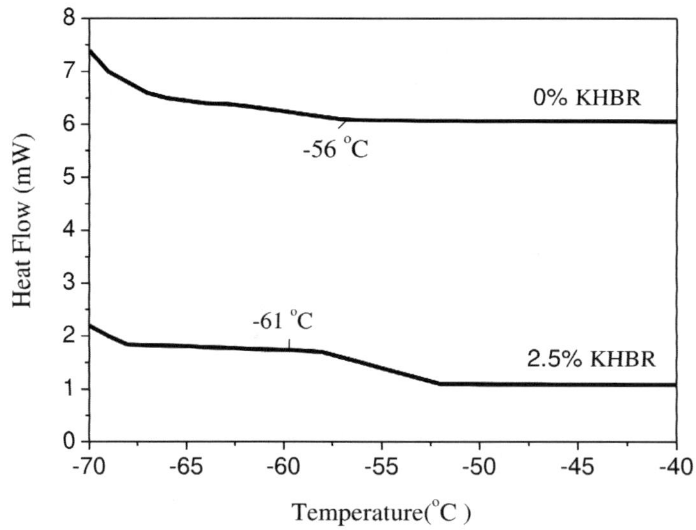

Figure 3. DSC thermograms of 0% HKBR and 2.5% HKBR.

3.5. DMA of KHBR-modified EPDM composites

The change of storage modulus and tan δ were shown in the Figure 4. The storage modulus was increased from 3300MPa to about 8300MPa with adding KHBR. The Tg, taken as the peak temperature of the tan δ curve, was only peak with 2.5% KHBR. However, the peak temperatures were not the case for 0% and 5% KHBR that was probably due to the inefficient network formation between KHBR and EPDM.

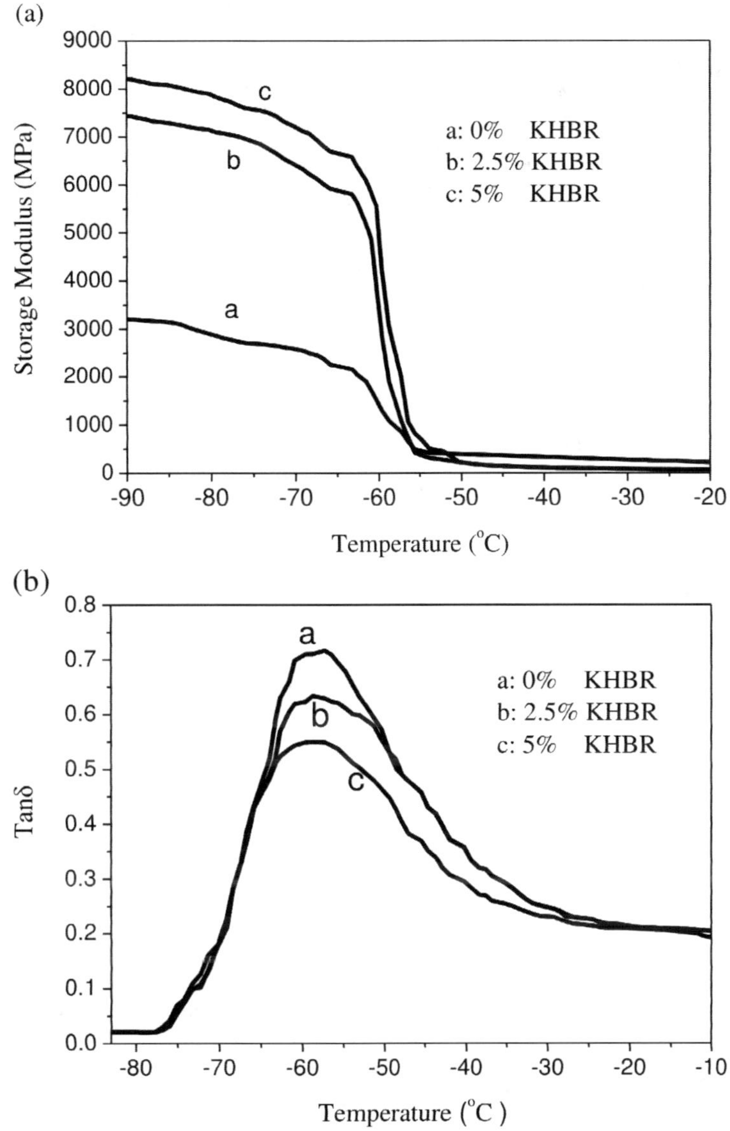

Figure 4. DMA test results (a) storage modulus and (b) tan δ .

4. Conclusions

EPDM composites modified with KH590-polybutadiene were studied in details, in which EPDM was used as a matrix, KHBR used as a compatibilizer. KH590 silane coupling plays an important key role in the rubber formula, the aim is to crosslink inorganic component. The addition of KHBR changed the mechanical and thermal properties of the EPDM. The mechanical properties improved with adding KHBR. EPDM rubber with adding 2.5% KHBR had the better mechanical properties. The results obtained from the MDR, DSC and kinetic studies were consistent and parallel.

Acknowledgments

This work was financially supported by Fundamental Research Funds in Heilongjiang Provincial Universities, China (No: YSTSXK201864).

IOP Conf. Series: Materials Science and Engineering **727** (2020) 012005 doi:10.1088/1757-899X/727/1/012005

References

[1] Weili W, He H and Zhe C 2019 *Sci. Eng. Compos. Mater.* **58** 209
[2] Seyed M, Ghasem N and Aghil B 2018 *Polym. Composite.* **39** 4071
[3] Amin A, Kandil H, Rabia M, Nashar E and Ismail N 2018 *Polym-Plast. Technol.* **57** 1733
[4] Weili W and He H 2019 *J. Polym. Sci. Pol. Phys.* **58** 330
[5] Dhruba J, Kinsuk N and Nikhil K 2013 *Eur. Polym. J.* **49** 4098
[6] Nicolas C, Oguzhan O, Edith PD, Bouvardj L, Pradille C and Billon N 2019 *Polymer.* **175** 329
[7] Benneghmouche Z and Benachour D 2019 *Compos. Interface.* **26** 711
[8] Kato A, Ikeda Y and Kohjiya S 2018 *Polym-Plast. Technol.* **57** 1418
[9] Karaagaç, Bagdagül, Kaner, Damla and Deniz 2010 *Polym. Composite.* **31** 1869
[10] Raut P, Swanson N, Kulkarni A, Pugh C and Janasanhan C 2018 *Polymer.* **148** 247

IOP Conf. Series: Materials Science and Engineering **727** (2020) 012006 doi:10.1088/1757-899X/727/1/012006

Effect of PS-g-MAH Content on ABS/PC Composites

Q Zhang [1,2,*], J Wen [1,2], B Chen[1,2], L Yin[1,2] and Z F Zhu[1,2]

[1] Jiangsu Xuzhou Construction Machinery Research Institute, Xuzhou 221004;
[2] State Key Laboratory of Intelligent Manufacturing of Advanced Construction Machinery, Xuzhou Construction Machinery Group, Xuzhou 221004

[*]Corresponding author: zhangqiong52806@163.com

Abstract. Acrylonitrile–Butadiene–Styrene (ABS) is a common plastic which used in many fields with two-phase system. Polycarbonate (PC) is a high-performance amorphous engineering thermoplastics. ABS/PC composites were researched in many academic papers and used in many fields. Polystyrene(PS) grafted with maleic anhydride (MAH)(PS-g-MAH) on ABS/PC composites is studied in this article. The tensile strength and elongation at break of ABS/PC blends have the maximum value when the content of PS-g-MAH is 3%. Tg of ABS in blends increase with the content of PS-g-MAH, which maybe induced by the increase of ABS/PC interface compatibility. From the photo of SEM, the addition of PS-g-MAH improves the compatibility of the interface of the two phases, leading to an increasing trend in the size of PC particles.

1. Introduction

Acrylonitrile–Butadiene–Styrene (ABS) is a popular plastic in industrial as well as commodity applications and being used in many fields like defence, aerospace, automobile, electrical, computer and telecommunications among the synthetic thermoplastics materials. ABS plastics are two-phase systems, and it is a hard and tough thermoplastic with good impact strength and surface gloss. Polycarbonate (PC) is a resin in which groups of dihydric or polyhydric phenols are linked through carbonate groups. PC is a high-performance amorphous engineering thermoplastics with an exceptionally high impact strength, clarity, heat resistance, dimensional stability, and a good electric property. The process ability of PC is rather difficult because its high melt viscosity hinders the fluidity of PC, and the residual stress resulting from processing causes fractures. To improve these problems, new polymer blends and alloys have been developed. ABS, PC and their alloys are widely used in automotive industry, computer and equipment housings. ABS/PC can be used to produce products with various physical properties and effective functions. Fillers modify the existing physicochemical characteristics but also invariably reduce cost, improve workability during processing, and develop new properties not present in the original resin [1-4].

PC / ABS blends have been investigated with respect to compatibility and mechanical behaviour aspects, processing conditions have a stronger effect on the phase morphology. The phase morphology of PC / ABS blends can be significantly affected by its processing conditions. A synergistic effect was observed for modulus of elasticity upon the addition of PC in ABS [5-7]. A compatibilizer named acrylonitrile – butadiene–styrene copolymer (ABS) grafted with maleic anhydride (MAH) (ABS- g-MAH) have good influence on the mechanical properties, and could improve the compatibility of the ABS/PC alloy significantly[8-10].

A compatibilizer named polystyrene (PS) grafted with maleic anhydride (MAH) (PS- g-MAH) was used in this article to improve the performance of ABS/PC alloy. The tensile strength, thermal property and micrograph of the samples were studied.

2. Materials

ABS, Basf co., LTD. PC, Basf co., LTD. PS- g-MAH, Xuzhou kebao experiment instrument co., LTD. Material proportions were listed in table 1.

Table 1 Material proportions

Number	ABS	PC	PS- g-MAH
1	100	40	0
2	100	40	3
3	100	40	6
4	100	40	9
5	100	40	12

2.1. Blends and sample preparation

2.1.1 Drying. ABS and PC were dried in an oven to remove absorbed water. The drying conditions were shown in Table 2.

Table 2 Drying conditions

material	Drying temperature(°C)	Drying time(hours)
ABS	80	12
PC	120	6

2.1.2 Blending and injection moulding. ABS/PC composites were prepared by melt mixing in a co-rotating twin-screw extruder, operating at screw speed of 210rpm with a temperature ranging from 190°C to 220°C which depending on the mixture composition. Then the particles of the blends were obtained from a knife pelletizer. ABS/PC blends were dried in a vacuum oven at 80°C for 6h, and injected from injection moulding machine with a temperature ranging 220 °C to 240°C.

2.2. Characterization

2.2.1 Test of mechanical properties. The tensile strength was tested by ETM104C electronic universal testing machine. Five samples of each composite were tested.

2.2.2 Test of glass transition temperature (Tg). Glass transition temperature (Tg) were tested by differential scanning calorimeter (DSC) (Perkin–Elmer DSC6000).

2.3. Morphology

The morphology of ABS/PC blends were tested by INSPECT S50 scanning electron microscopy (SEM) after the composites were covered with a thin layer of gold.

3. Results and discussion

3.1 Tensile properties

Figure 1 shows the tensile strength and the elongation at break for different contents of PS- g-MAH. The tensile strength and elongation at break of ABS/PC blends tend to increase first and then decrease with the addition of PS- g-MAH. The maximum value is 45.12MPa, 94% when the content of PS- g-MAH is 3%. The carboxyl group in PS- g-MAH can increase the compatibility between ABS and PC interface. There is an excellent compatibility between PS- g-MAH and ABS, which induce the strength of ABS/PC blends increases. However, the tensile strength decreased as the content of PS- g-MAH continued to increase. That may be induced by the excessive content of PS- g-MAH, which has become a three-phase component of ABS/PC/PS rather than just a compatibilizer. PS and PC are incompatible phases with each other, which may reduce tensile properties.

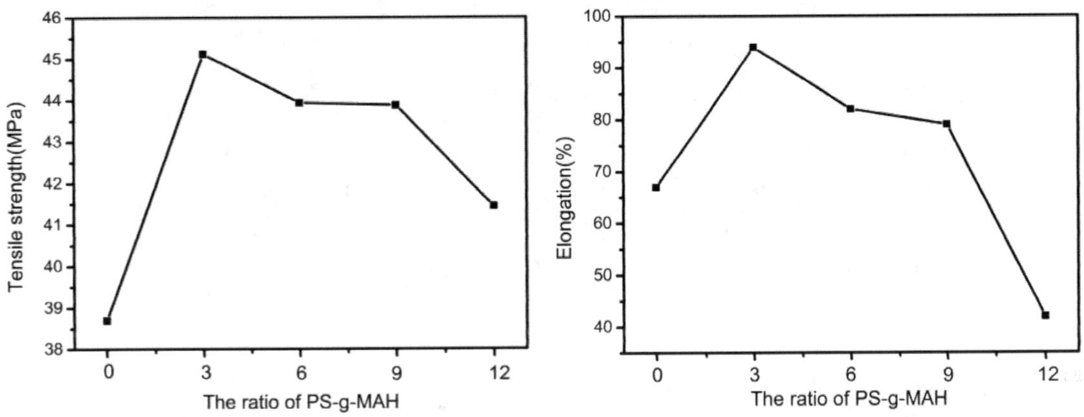

(a) The tensile strength of ABS/PC blends (b) The elongation at break of ABS/PC blends

Figure 1. The tensile properties of ABS/PC blends with different content of PS- g-MAH

3.2 Thermal property

Glass transition temperature (Tg) of ABS in blends are shown in figure 2. It can be seen that Tg of ABS shift with the increase of PS- g-MAH. Tg of ABS reaches 104.35°Cwhen the content of PS- g-MAH reaches 12%. There are two reasons for the increase of the glass transition temperature of ABS. First, the addition of PS-g-MAH increases the interface compatibility of ABS and PC, then decline the interface energy. Second, the increase of PS-g-MAH content will increase the proportion of PS components in ABS, which could be change the glass transition temperature of ABS in blends. That maybe induced by the increase of ABS/PC interface compatibility.

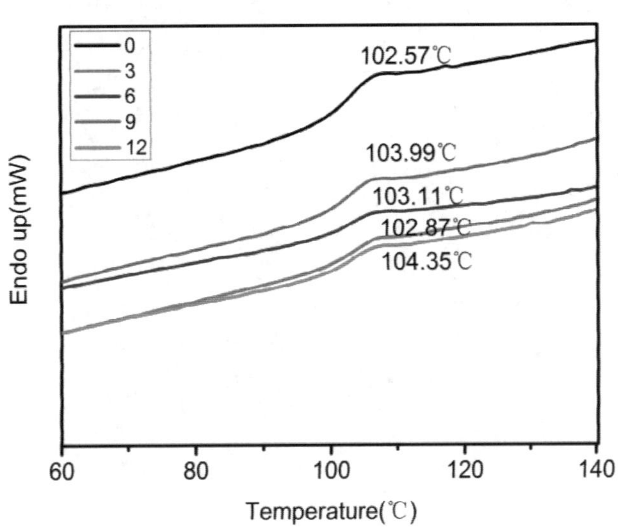

Figure 2. Tg of ABS in ABS/PC blends with different content of PS-g-MAH

IOP Conf. Series: Materials Science and Engineering **727** (2020) 012006 doi:10.1088/1757-899X/727/1/012006

3.3Morphology

The morphology of ABS/PC blends are shown in figure 3. It can be clearly seen that ABS and PC are incompatible phases. And PC is dispersed in the ABS matrix evenly in globular shape. The addition of PS-g-MAH cannot integrate ABS and PC perfectly. We also observed that the size of PC particles dispersed in ABS matrix gradually increased with the increase of PS-g-MAH content from figure 3, table 3. This may be because the addition of carboxyl groups in maleic anhydride improves the interface compatibility of PC and ABS and reduces the interface energy.

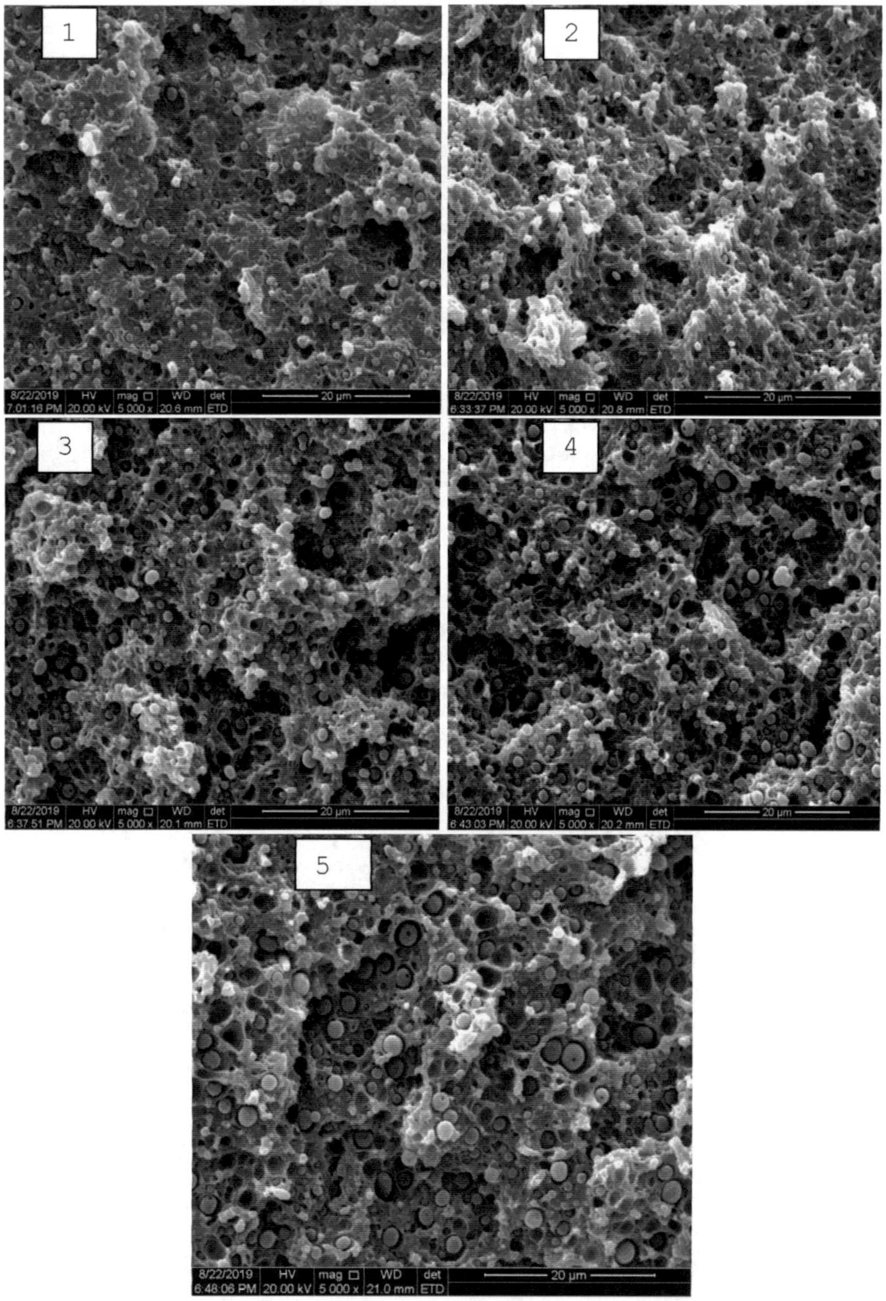

Figure 3. The morphology of ABS/PC blends with different content of PS-g-MAH 1.0phr, 2.3phr, 3.6phr, 4.9phr, 5. 12phr

IOP Conf. Series: Materials Science and Engineering **727** (2020) 012006 doi:10.1088/1757-899X/727/1/012006

Table 3 Diameter of PC particle

The content of PS-g-MAH	Diameter of PC particle
0	1.07
3	1.34
6	1.46
9	1.73
12	1.81

4. Conclusion

The tensile strength and elongation at break of ABS/PC blends increase first and then decrease with the addition of PS-g-MAH. The maximum value occurs when the content of PS-g-MAH is 3%.

Tg of ABS increase with the content of PS-g-MAH, that may be because PS-g-MAH improve the interface compatibility of ABS and PC, then reduce the interface energy.

ABS and PC are incompatible phases, the addition of PS-g-MAH improves the compatibility of the interface of the two phases, leading to an increasing trend in the size of PC particles.

References

[1] H.J.CHOI,S.H.PARK,J.K.KIM,J.I.JUN 2000 *J.Appl.Polym.Sci.* **75**,417
[2] B.Perret, K.H.Pawlowski, B.Schartel 2009 *J Therm Anal Calorim* **97**,949
[3] Y.Sun,Z.X.Guo,J,Yu 2010 *Macromol.Mater.Eng.* **295**,263
[4] M.I.Triantou,P.A.Tarantili 2012 *Macromol.Symp.* 321-322,155
[5] B.I.Chaudhry,E.Hage,L.A.Pessan 1998 *J.Appl.Polym.Sci.* **67**,1605
[6] M.M.K.Khan, C.J.Hilado,S.Agarwal,R.K.Gupta 2007 *J Polym Environ* **15**,188
[7] A.Guinault,C.Sollogoub,Int 2009 *J Mater Form* **2**,701
[8] X.F.Zhang,Y.S.Chen,Y.Zhang,Z.L.Peng,Y.X.Zhang,W.Zhou 2001 *J.Appl.Polym.Sci.* **81**,831
[9] J.Zhang 2016 *Procedia Eng* **135**,83
[10] M.I.Triantou,K.I.Stathi,P.A.Tarantili 2019 *Polym Composite* **10**,1662

Optimization of PC-GTAW Orbital Welding Parameters of AISI 304L Stainless Steel Pipe Using ANOVA and Taguchi Method

A S Baskoro[1,*], G Kiswanto[1] and A Widyianto[1]

[1] Mechanical Engineering Department, Faculty of Engineering Universitas Indonesia, Kampus Baru UI Depok, 16424, Indonesia

[*] E-mail: ario@eng.ui.ac.id

Abstract. In the present study, the PC-GTAW orbital welding parameters of AISI 304L stainless steel pipe have been optimized by using ANOVA and Taguchi method. The welding method uses Pulsatile Current-GTAW without added material (autogenous). Peak currents were varied in this study and had the same average current of 100 ± 0.5 amperes. The Taguchi method was applied to search for OBW, DoP and UTS that are optimal and L_{16} Taguchi's orthogonal array with two factors, four levels and two replications. S/N ratio with a quality character of larger is better (LB) was used to find the maximum value. The results show that raising the peak current would widen the OBW and deepen the DoP but reduce UTS. The 35-C parameter was the optimal parameter to produce OBW and DoP while the 100-C parameter was the optimal parameter to produce maximum UTS.

Keyword: Orbital welding, PC-GTAW, Taguchi, AISI 304L

1. Introduction

The pipe is one of the most functional materials as an effective means of transportation such as water, gas, oil, etc. One type of material commonly used is stainless steel. Stainless steel has the characteristics of corrosion resistance at high temperatures and pressures [1]. One of the welding methods used is orbital pipe welding. This welding has disadvantages in the bead width and depth of penetration of welding results due to the influence of the gravitational force. To minimize the effects of gravitational forces can use the pulsating current welding method [2, 3]. Optimization of welding parameters is needed for orbital pipe welding to get the best weld results.

The previous researcher examined that orbital pipe welding is affected by the welding position and welding current [3]. Joseph [4] reported the results of his research on the optimization of GTAW welding parameters pulsating current using simulated annealing and genetic algorithm, that the peak current, base current, welding speed and gas flow rate had an influence on tensile strength and elongation. The Taguchi method can be used to find optimal parameters to improve the tensile strength of S30430 steel plate [5] and API 5L X65 pipes [6]. Ragavendran [7] successfully optimizes hybrid laser welding parameters - TIG uses RSM in 316LN material.

Based on studies that have been carried out that optimization of welding parameters is still needed to improve weld quality. Therefore this paper aims to find the optimal parameters to obtain the best tensile strength in orbital pipe welding. The material used in this study is AISI 304L stainless steel pipe with pulsatile current welding for the welding method. The optimization method used is the Taguchi method with ANOVA analysis.

2. Experimental Methods

2.1 Material

The material used in this study was AISI 304L stainless steel pipe with outer diameter 114.2 mm and thickness of 3 mm. The optical emission spectrometer (OEM) was used to test the chemical composition of AISI 304L stainless steel pipe and the chemical composition was listed in Table 1. Before welding, the surfaces of all specimens were cleaned by 80 until 400 grit flexible sandpapers. After that, it was cleaned with acetone solution to remove surface impurities.

Table 1. Chemical composition (wt %) of AISI 304L pipe

	C	Si	Mn	P	S	Cr	Mo
AISI 304L	0.026	0.400	1.44	0.037	0.011	19.5	0.118
	Ni	Al	Cu	Nb	Ti	V	Fe
	8.26	0.006	0.096	0.018	0.002	0.078	bal.

2.2 Experimental design and optimization

The welding method used in this study were without filler metal (autogenous) and using orbital pipe welding. A GeKaMac power TIG 2200 DC pulse welding machine was used in this study with 1 phase input voltage, 230 V, 50-60 Hz and current range 5-220 A. The top weld zone was shielded by high purity argon (99.99%) gas with a flow rate of 11 L/min. A tungsten electrode was used EWTh-2 with a red code AWS with diameter 2.4 mm and sharpening angle of 30^0. The welding speed was set constant 1.40 mm/s. The welding parameters and the process parameters that were used in this study were shown in Table 2 and Table 3, respectively.

Table 2. Pulsatile current GTAW parameters

No	Code	Peak current (A)	Base current (A)	Peak time (ms)	Base time (ms)	Frequency (Hz)	Duty cycle (%)	Mean current (A)	Welding speed (mm/s)
1	35-C	212	40	70	130	5	35	100.2	1.4
2	50-C	160	40	100	100	5	50	100	1.4
3	65-C	133	40	130	70	5	65	100.45	1.4
4	100-C	100	100	100	100	5	100	100	1.4

Table 3. Process parameters at four level

No	Parameters	Level 1	Level 2	Level 3	Level 4
1	Welding current (code)	35-C	50-C	65-C	100-C
2	Welding position (degree)	0	90	180	270

A pulsatile current GTAW (PC-GTAW) method was adopted, with variations of peak currents, base currents and duty cycles. In pulsatile currents GTAW has the same mean current of 100 ± 0.5 Amperes and the average excitation voltage during the welding process was between 12–13 volts. Furthermore, the heat input calculation was received by the material in pulsatile current welding. Heat input in the pulsatile current GTAW can be calculated from the mean current. Control of heat input can be made with adjustments peak current, base current and duty cycle [8]. Figure 1 shows the

IOP Conf. Series: Materials Science and Engineering **727** (2020) 012007 doi:10.1088/1757-899X/727/1/012007

welding results on the welded pipe surface using one of the pulsed welding parameters. The pipe was cut into four parts then OBW was seen visually after that measurements were made using a digital microscope. For DoP measurements, it has been prepared macrostructure previously. While for tensile testing carried out by ASTM E-8M standard.

Figure 1. The weld surface profile of pipe

The Taguchi technique was applied to conduct the optimization of PC-GTAW parameters. The experimental procedure included the design of experiment, materials, and welding parameters. The Taguchi method can study data with minimum experimental runs. Based on the Taguchi method, (L_{16}) orthogonal arrays with four levels, two factors, and two replications were employed as presented in Table 4.

Table 4. Taguchi's L_{16} orthogonal array

Experiment number	Process parameters		Responses					
	Welding current (code)	Welding position (degree)	OBW (mm)		DoP (mm)		UTS (MPa)	
			Rep. I	Rep. II	Rep. I	Rep. II	Rep. I	Rep. II
1	35-C	0	7.76	7.46	0.91	1.25	613.24	647.86
2	35-C	90	7.44	8.01	1.15	1.18	678.41	648.32
3	35-C	180	7.63	8.20	0.56	0.85	684.34	667.43
4	35-C	270	7.42	7.93	0.84	0.54	496.83	545.34
5	50-C	0	6.44	6.64	0.31	0.17	653.61	620.47
6	50-C	90	6.79	7.56	1.28	0.57	593.88	604.56
7	50-C	180	7.11	7.42	0.59	0.79	613.51	647.24
8	50-C	270	8.32	7.23	0.07	0.05	626.80	640.89
9	65-C	0	7.24	7.39	0.37	0.24	611.66	645.67
10	65-C	90	6.97	5.92	0.11	0.19	602.33	656.24
11	65-C	180	6.53	6.55	0.07	0.19	664.78	638.57
12	65-C	270	6.49	6.87	0.11	0.23	670.67	659.86
13	100-C	0	6.65	7.32	0.38	0.10	637.27	715.22
14	100-C	90	8.15	6.91	0.28	0.15	683.08	655.55
15	100-C	180	6.02	6.32	0.04	0.08	636.35	649.43
16	100-C	270	6.89	6.59	0.17	0.10	659.81	663.50

Taguchi uses a signal-to-noise (S/N) ratio to find the optimal value (maximum and minimum) of several factors used. The S/N ratio depends on several types of quality characteristics, it consists of 4 types defined as smaller is better (SB), nominal is best (NB), larger is better (LB), and fraction defective (FD) [9, 10]. In this study, to search the maximum value of OBW, DoP and UTS using the larger is better method. The S/N ration for LB can be calculated using the following Equation [6].

$$S/_N \text{ ratio} = -10\log\left(\frac{1}{n}\sum\frac{1}{y_i^{\,2}}\right) \quad (1)$$

In additional, ANOVA was applied to determine the significance of each input parameters against the response parameters.

3. Results and Discussion

3.1. Analysis of outer bead width (OBW)

The ANOVA analysis for OBW was shown in Table 5. Welding currents have a very high contribution to the formation of OBW by 93.54% while the welding position only contributes of 6.46%. This is confirmed by the result of the response from OBW as shown in Table 6. Increasing the peak current in welding with pulsatile currents can widen the OBW while the welding position has little effect on the OBW.

Table 5. ANOVA analysis for outer bead width

Source	DF	Sum of squares	Mean of squares	F-value	p-value	P.C.
Welding position	3	0.1626	0.05419	0.23	0.870	6.46
Welding current	3	2.3525	0.78416	3.40	0.067	93.54
Error	9	2.0758	0.23064			
Total	15	4.5908				

Table 6. Response table for OBW, DoP and UTS

OBW			DoP			UTS		
Level	Welding position	Welding current	Level	Welding position	Welding current	Level	Welding position	Welding current
1	17.02	17.75	1	-10.342	-1.442	1	56.14	55.83
2	17.09	17.09	2	-8.329	-11.120	2	56.11	55.91
3	16.82	17.54	3	-13.412	-16.506	3	56.25	56.16
4	17.12	16.67	4	-15.995	-19.010	4	55.80	56.41
Delta	0.30	1.21	Delta	7.667	17.569	Delta	0.45	0.58
Rank	2	1	Rank	2	1	Rank	2	1

3.2. Analysis of depth of penetration (DoP)

The ANOVA analysis for DoP was shown in Table 7. Welding currents have a significant contribution to the formation of DoP by 85.04% while the welding position only contributes of 14.96%. This is confirmed by the result of the response from DoP as shown in Table 6. DoP in orbital pipe welding is strongly influenced by welding current other than the effect of the welding position and the gravitational force.

Table 7. ANOVA analysis for depth of penetration

Source	DF	Sum of squares	Mean of squares	F-value	p-value	P.C.
Welding position	3	0.2550	0.08501	1.72	0.232	14.96
Welding current	3	1.4497	0.48325	9.79	0.003	85.04
Error	9	0.4443	0.04937			
Total	15	2.1491				

3.3. Analysis of ultimate tensile strength (UTS)

The ANOVA analysis for UTS was shown in Table 8. The welding position and welding current have a contribution of 32.05% and 67.95%, respectively to UTS. The welding current still has a higher contribution than the welding position as shown in Table 6. Increasing the peak current in welding with pulsatile currents has little effect on UTS.

Table 8. ANOVA analysis for ultimate tensile strength

Source	DF	Sum of squares	Mean of squares	F-value	p-value	P.C.
Welding position	3	1948	649.3	0.38	0.771	32.05
Welding current	3	4130	1376.7	0.80	0.523	67.95
Error	9	15421	1713.5			
Total	15	21499				

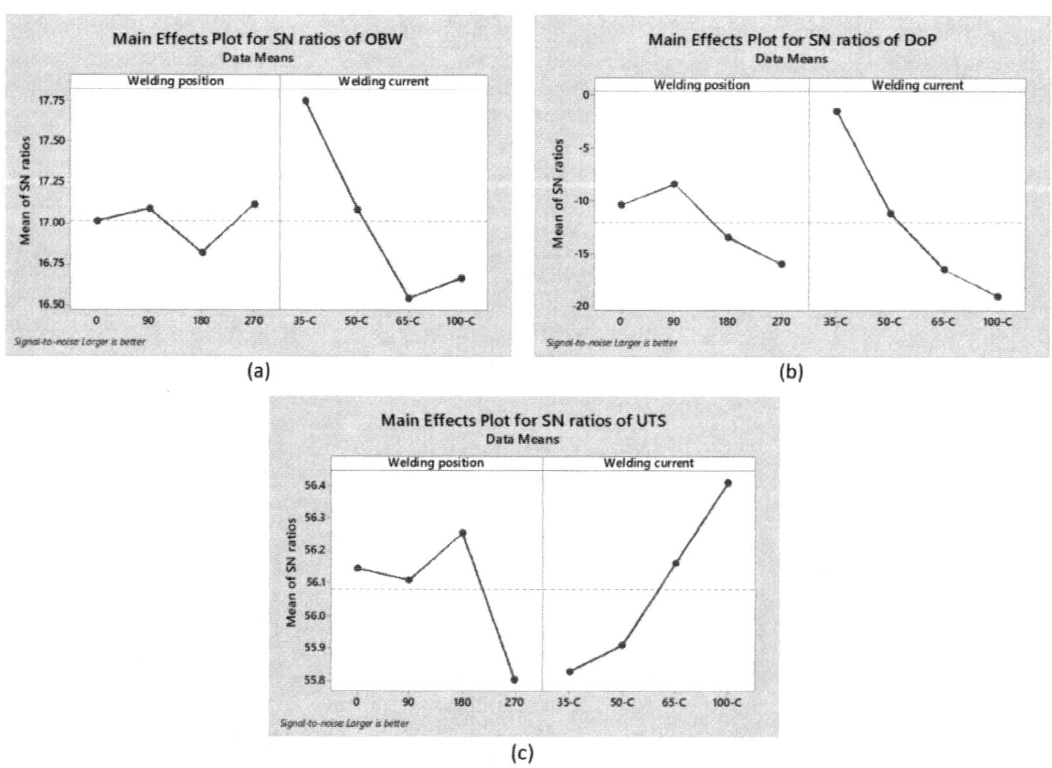

Figure 2. Main effect plot for SN ratios of (a) OBW, (b) DoP and (c) UTS

Main effects SN ratio of OBW, DoP and UTS were plot in Figure 2a-c. Raising the peak current in welding with pulsatile current can increasing the OBW and DoP but decreasing the UTS. In parameter 35-C has the maximum OBW and DoP but produces the smallest UTS. The 35-C parameter produces optimal OBW and DoP with welding positions 270^0 and 90^0, respectively. The optimal parameter for producing maximum UTS was 100-C as welding current and 180^0 as a welding position with sufficient OBW and DoP.

4. Conclusions

In general, raising the peak current in pulsatile currents welding would increase OBW and DoP for several welding positions, but conversely the resulting UTS would decrease. Indirectly OBW and DoP have a correlation to the UTS produced even though there were other factors that influence it. In parameter 35-C has the maximum OBW and DoP with welding position at 270^0 and 90^0, respectively but produces the smallest UTS. While the 100-C parameter and welding position at 180^0 produces the most maximum UTS with the appropriate OBW and DoP. The welding current has a large contribution of 65% - 95% while the welding position only contributes 6% - 35% to get the maximum OBW, DoP and UTS.

Acknowledgments

This research is supported by the Master Program to Doctorate for Scholar Excellent (PMDSU) program of the Ministry of Research & Technology and High Education (RISTEK DIKTI) 2019 with contract number NKB-1855/UN2.R3.1/HKP.05.00/2019.

References

[1] J. Xu, J. Chen, Y. Duan, C. Yu, J. Chen, and H. Lu, "Comparison of residual stress induced by TIG and LBW in girth weld of AISI 304 stainless steel pipes," *Journal of Materials Processing Technology*. 248 pp. 178-184, 2017.

[2] H. Eisazadeh, D. J. Haines, and M. Torabizadeh, "Effects of gravity on mechanical properties of GTA welded joints," *Journal of Materials Processing Technology*. 214 (5) pp. 1136-1142, 2014.

[3] D. W. Figueirôa, I. O. Pigozzo, R. H. G. e. Silva, T. F. d. A. Santos, and S. L. Urtiga Filho, "Influence of welding position and parameters in orbital tig welding applied to low-carbon steel pipes," *Welding international*. 31 (8) pp. 583-590, 2017.

[4] J. Joseph and S. Muthukumaran, "Optimization of pulsed current GTAW process parameters for sintered hot forged AISI 4135 P/M steel welds by simulated annealing and genetic algorithm," *Journal of Mechanical Science and Technology*. **30** (1) pp. 145-155, 2016.

[5] A. Ahmad and S. Alam, "Grey Based Taguchi Method for Optimization of TIG Process Parameter in Improving Tensile Strength of S30430 Stainless Steel," in *IOP Conference Series: Materials Science and Engineering*, 2018, vol. 404, no. 1, p. 012003: IOP Publishing.

[6] H. Ada and C. Çetinkaya, "Optimization of maximum tensile strength of welded joints of API 5L X65 pipes by Taguchi method," *Materials Research Express*. **6** (3) p. 036526, 2018.

[7] M. Ragavendran, N. Chandrasekhar, R. Ravikumar, R. Saxena, M. Vasudevan, and A. Bhaduri, "Optimization of hybrid laser–TIG welding of 316LN steel using response surface methodology (RSM)," *Optics and Lasers in Engineering*. 94 pp. 27-36, 2017.

[8] T. S. Kumar, V. Balasubramanian, and M. Sanavullah, "Influences of pulsed current tungsten inert gas welding parameters on the tensile properties of AA 6061 aluminium alloy," *Materials & design*. 28 (7) pp. 2080-2092, 2007.

[9] S. Pandiarajan, S. S. Kumaran, L. Kumaraswamidhas, and R. Saravanan, "Interfacial microstructure and optimization of friction welding by Taguchi and ANOVA method on SA 213 tube to SA 387 tube plate without backing block using an external tool," *Journal of Alloys and Compounds*. 654 pp. 534-545, 2016.

[10] K. Krishnaiah and P. Shahabudeen, *Applied design of experiments and Taguchi methods*. PHI Learning Pvt. Ltd., 2012.

IOP Conf. Series: Materials Science and Engineering **727** (2020) 012008 doi:10.1088/1757-899X/727/1/012008

Influence of Tools Diameters and Plunge Depth on Mechanical Properties of Micro Friction Stir Spot Welding Materials A1100

A.S. Baskoro*, A.Riyanto, M.F. Arifardi and P.Rupajati

Mechanical Engineering Department, Faculty of Engineering, Universitas Indonesia, Depok, Indonesia

*Corresponding author's e-mail address: ario@eng.ui.ac.id

Abstract. Welding on thin aluminum alloy material is still relatively difficult, so that appropriate welding technology is needed to joint the ultra-thin aluminum material. One of the processing welding capable of solving to ultra-thin aluminium is micro friction stir spot welding (μFSSW). The micro FSSW process is an adaptation of the FSW process where solid-state processes allow the welding process without changes in the basic characteristics of the workpiece. The purpose of this study was to analyze the effect parameters used in this study are pinless tool diameter and plunge depth on micro FSSW. The material used in this study was A1100 with a thickness of 0.42 mm using lap joint. The welding parameters used in this study are tool diameter and plunge depth, where each parameter has three levels. While the response variables observed were mechanical and fractures properties. To analyze maximum shear strength/load using response surface methodology (RSM). Experimental results and analysis of influence tool diameter and plunge depth on micro-friction stir spot welding (μFSSW) analysis show that tool diameter and plunge depth are directly proportional to maximum shear load. Furthermore, the smallest hook defect occurs on the tool diameter 5 mm with plunge depth 500 μm.

Keyword: FSSW, plunge depth, A1100, RSM

1. Introduction

Nowadays, it was found that one of the welding technology is welding that utilizes heat over the friction that occurs, namely Friction Stir Welding (FSW). This technology was discovered by The Welding Institute in the United Kingdom in 1991 [1]. FSW provides several metallurgical, environmental and energy-saving advantages. The solid-state process produces friction heat that is not too high so it is suitable to be applied to aluminum materials that have high heat transfer properties, which are difficult to control by the welding method with fusion in general. Several studies on welding micro friction stir spot welding on material aluminum alloys have developed [2, 3]. The FSW welding process produces a stable joining dimension and low distortion. Micro friction stir welding (μFSW) is an adaptation of FSW in materials with a thickness of less than 1 mm. μFSW is an adaptation of FSW in materials with a thickness of less than 1 mm. NJ Teh et al. [4] have developed a micro-application of FSW in a journal published by TWI, discussing issues about challenges, tools, and downscaling machines for μFSW, and the results obtained. Yin, et al. [5] conducted an FSW application study on

Spot Welding, resulting in friction stir spot welding. Further development of the application on the micro-scale, it became a micro-friction stir spot welding by Wang, et al. [6]. To get good quality and joint strength is not easy, because the welding process parameters must be adjusted to the joining method and the material. These parameters differ from one another so that it cannot be applied to all materials. Tool geometry and plunge depth in the μFSSW welding process is one of the most critical parameters. Research on welding parameters in other μFSSWs such as the effect of dwell time and plunge speed during the μFSSW process on aluminum A1100 thin plates [7] and also the effect of tool rotational speed has been carried out by Baskoro et al. [8]. Some research investigated friction stir spot welding using a tool without pin. Tozaki et al [9] investigated friction stir spot welding A6061-T4 using a tool without pin. In this study, the influence of two welding parameters (tool diameter and plunge depth) on mechanical properties and fracture properties of the joints micro-friction stir spot welding of thin material A1100 are investigated.

2. Experimental Procedures

In this study, thin aluminum plates A1100 with a dimension of 2.5 mm x 6 mm x 0.42 mm were friction stir spot lap welded, as shown in figure 1a. The chemical composition of the aluminum alloy 1100 is 99.00% min Al, 0.05-0.20% of Cu, 0.95% max of Si+Fe, 0.05% max of Mn, 0.10 % max. The hardness of A1100 is 28 HRB, ultimate tensile strength is 95-130 Mpa, elongation is 3-10%, yield strength is 105 MPa, the elastic modulus is 70-80 GPa[10]. The A1100 is pure aluminum used in manufacturing industries such as chemical plant equipment, food packaging industry, reflector lamps, cable sheaths. Tools are the most important part of μFSSW. In this study, the tool designed is pinless tools. The diameter of the tool is varied between; 2 mm, 4 mm and 6 mm, as shown in figure 1b. The tool material used is high speed steel (HSS) which is one variant of tool steel. The selected of this type of material is in accordance with the consideration of standards, availability of material, physical properties, capable of the engine, and the lifetime or toughness of the material. There are two microfriction stir spot welding parameters are tool diameter and plunge depth. The dwell time used as constant variable is 4 s. As shown in table 1, there are three levels for each parameter.

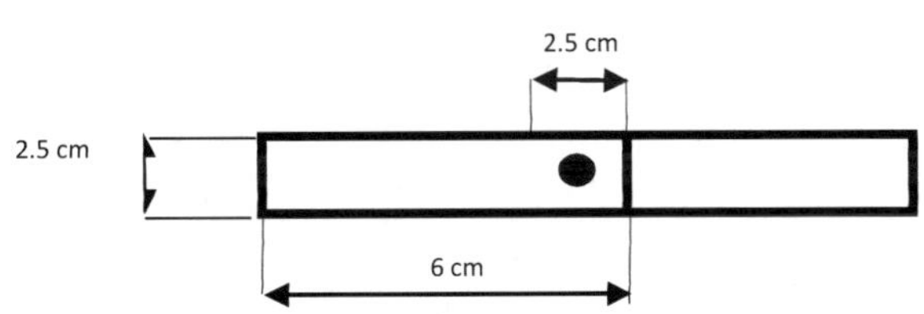

(a)

(b)

Figure 1. (a) The Schematic illustration of lap shear specimen; (b) variation of diameter tools

Table 1. Welding Parameters and their levels

No	Parameter	unit	Notation	Level		
				-1	0	1
1	Tool Diameter	mm	A	2	4	6
2	Plunge Depth	μm	B	300	400	500

3. Result and Discussion

3.1. Shear Tensile Strength

The results of maximum shear load and maximum shear strength, as shown in table 2. This test was conducted were nine variations, each variation was done three replications and led to a total of 27 experiments.

Table 2. The result of maximum shear load and maximum shear strength

No	Coded Value A	Coded Value B	Tools Dia (mm)	Insert Depth (μm)	Max Shear Load (N)	Joint Dia (mm)	Thick Joint (mm)	Joint Area (mm^2)	Max Shear Strength (MPa)
1	-1	-1	2	300	163.6	2.63	0.4	3.3	49.5
2	-1	-1	2	300	166.4	2.63	0.4	3.3	50.4
3	-1	-1	2	300	170.9	2.63	0.4	3.3	51.7
4	-1	0	2	400	241.3	2.67	0.4	3.4	72
5	-1	0	2	400	231.5	2.67	0.4	3.4	69
6	-1	0	2	400	206.7	2.67	0.4	3.4	61.6
7	-1	1	2	500	281.9	2.6	0.4	3.3	86.3
8	-1	1	2	500	251.9	2.6	0.4	3.3	77.1
9	-1	1	2	500	262.3	2.6	0.4	3.3	80.3
10	0	-1	4	300	313.4	5.2	0.4	6.5	48
11	0	-1	4	300	322.2	5.2	0.4	6.5	49.3
12	0	-1	4	300	365.3	5.2	0.4	6.5	55.9
13	0	0	4	400	368.5	5.17	0.4	6.5	56.7
14	0	0	4	400	340.9	5.17	0.4	6.5	52.5
15	0	0	4	400	398.5	5.17	0.4	6.5	61.4
16	0	1	4	500	474.8	5.17	0.4	6.5	73.1
17	0	1	4	500	515.7	5.17	0.4	6.5	79.4
18	0	1	4	500	498	5.17	0.4	6.5	76.7
19	1	-1	6	300	375.3	4	0.4	5	74.7
20	1	-1	6	300	375.3	4	0.4	5	74.7
21	1	-1	6	300	370.6	4	0.4	5	73.8
22	1	0	6	400	431.9	4.3	0.4	5.4	80
23	1	0	6	400	435.7	4.3	0.4	5.4	80.7
24	1	0	6	400	425.1	4.3	0.4	5.4	78.7
25	1	1	6	500	502.3	4.8	0.4	6	83.3
26	1	1	6	500	546.4	4.8	0.4	6	90.6
27	1	1	6	500	533	4.8	0.4	6	88.4

Based on table 2, there is a variation of the maximum shear load and shear strength values obtained. Table 2 shows that the maximum shear load increases every increase in plunge depth used. An increase in the tool diameter width also results in an increase in the maximum shear load. In samples with tool diameters of 4 mm and 6 mm and 500 μm of plunge depth have a maximum shear load that is close to each other. Meanwhile, the largest displacement is obtained from the shear test results of samples with an input diameter of 4 mm and plunge depth of 500 μm, where the displacement is more than 3 mm and still continues. Displacement of other samples is in the range of 1 mm. Meanwhile, the last two sample groups were sample groups with input diameter of 2 mm with a plunge depth of 400 μm and 300 μm having the lowest shear resistance in the range of numbers below 250 N, with large displacements in the range below 0.7 mm. The maximum shear load results from μFSSW (Y) are a function of variations in tool diameter parameters (A), and insert depth (B), and can be stated as follows equation (1).

$$Y = f(A, B) \quad (1)$$

The secondary order polynomial (regression) equation used to represent the Y response is as follows equation (2) :

$$Y = k_0 + \sum_{i=1}^{n} ki.xi \; , \; \sum_{i=1}^{n} ki^2.xii + \sum_{i=1}^{n} kij.xi.xj \quad (2)$$

Hence the equation obtained from the three parameters used in μFSSW welding is as follows equation (3):

$$Y = k_0 + k_1 A + k_2 B + k_3 A^2 + k_4 AB + k_5 B^2 \quad (3)$$

Where: k_0 is the average of the response, while k_1, k_2….. k_5 are regression coefficients. The values of k_1, k_2…..k_5 can be searched using the Matlab software so that they are obtained equation (4):

$$Y = 387.5 + 112.2A + 69.08B + -67.92A^2 + 13.78\ AB + 18.28B^2 \quad (4)$$

From the coefficient values obtained from calculations using Matlab, it can be seen that the welding variables (A and B) have an influence on the response (y) for the maximum shear load. In both of these equations, it can be seen that variable A, which is the diameter of the tool has a greater coefficient compared to variable B, namely the plunge depth. This shows that the tool diameter variable has a more dominant influence than the plunge depth variable. From Equation 4 and the response value obtained from the measurement, a response surface graph can be generated which describes the relationship between tool diameter variables and plunge depth to the maximum shear load.

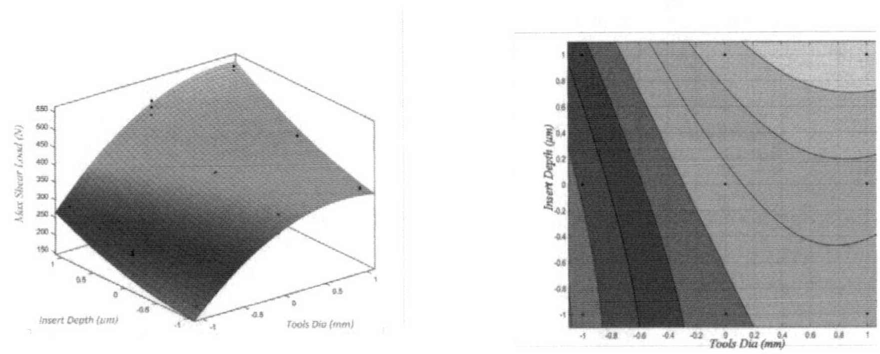

(a) (b)

Figure 2. a). RSM graph of relationship between tool diameter and plunge depth for maximum shear load; b). The contour of the relationship between tool diameter and plunge depth for maximum shear load

Figure 2a, it can be seen that the greater the diameter of the welding tool, the maximum shear load increase. Likewise with plunge depth if the greater the plunge depth, the maximum shear load increase. In both graphs, it also shows that the tool diameter has a more dominant relationship to the maximum shear load. Figure 2b shows that the influence of the plunge depth is more significant at the larger plunge depth. It can be seen that the graph trend is increasing with an increase in the plunge depth and the tool diameter.

3.2. Fracture properties

Fracture analysis with a width of 2 mm tool diameter is a type of shear rupture. In samples a 2 mm tool diameter and 500 µm plunge depth, fracture occurs at the top of the joint, as shown in figure 3a. In addition, there are also cracks in that section.

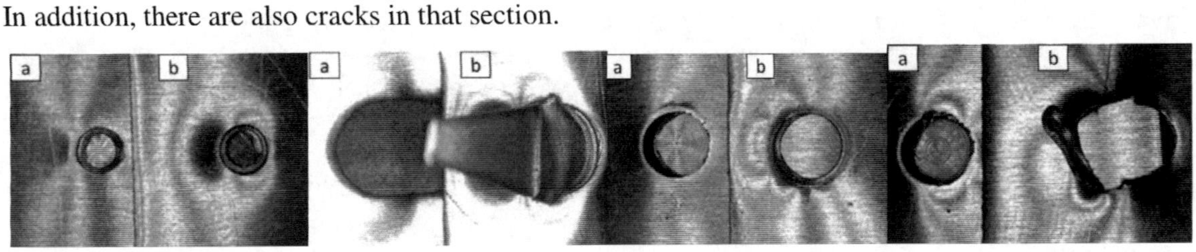

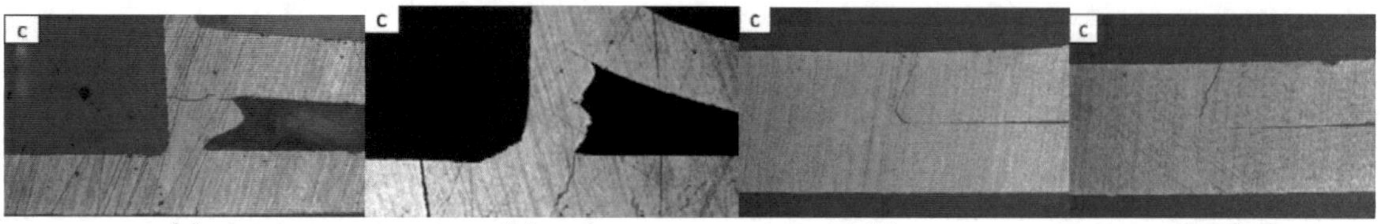

(a) (b) (c) (d)

Figure 3. Analysis and prediction of fracture in samples with tool diameter and plunge depth:(a) 2 mm and 500 μm; (b) 4 mm and 500 μm; (c) 6 mm and 400 μm; (d) 6 mm and 500 μm.

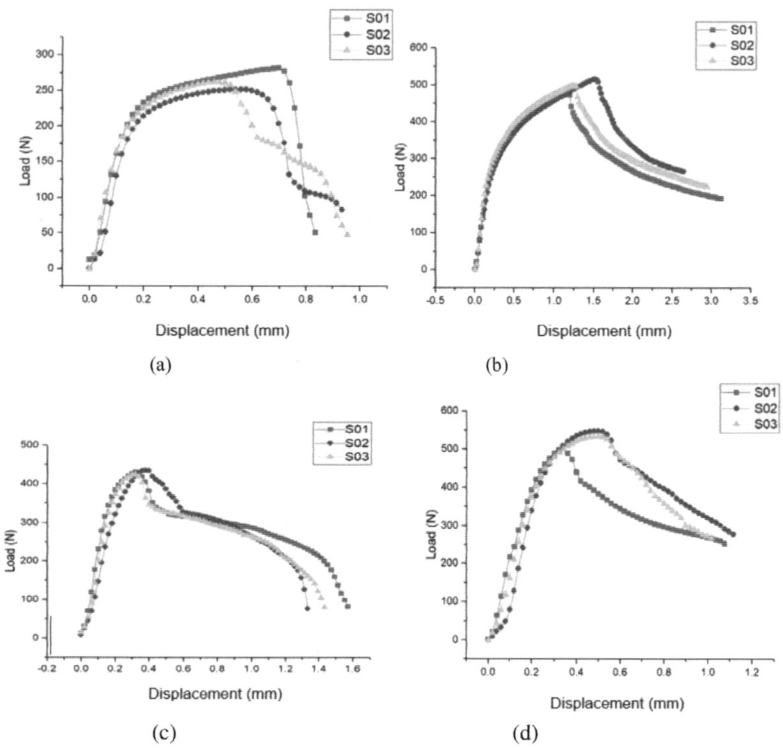

Figure 4. Load vs displacement for samples with tool diameter and plunge depth (a) 2 mm and 500 μm; (b) 4 mm and 500 μm; (c) 6 mm and 400 μm; (d) 6 mm and 500 μm

Figure 3b shows the type of shear rupture that also occurs in sample 4 mm tools and 500 μm a plunge depth of are visible fractures and tear in the base material. This fracture is usually called base plate tearing. Figure 3c and 3d show the type of fracture which occurs in the sample with 6 mm diameter tools are a type of rupture plug fracture. The fracture in the sample with a 6 mm diameter tool and a 500 μm plunge depth showed fractures that extend from the weld joint to the base metal around the weld, as shown in figure 3d. While at a depth of 400 microns, plug rupture occurs only around the weld diameter area, as shown in figure 3c. Based on micro testing observations, visible cracks that form a basin upward or called a hook defect [11]. Hence, the greater the plunge depth, the shorter the crack occurs. However, at a plunge depth of 500 μm, it may produce an optimum joining. From the results of the tensile test which is displayed in table 2 and figure 4 indicated the greater the plunge depth produces greater maximum shear load.

4. Conclusion

In this study, the influence of tools diameter and plunge depth on mechanical properties thin material A1100 using micro-friction stir spot welding was analyzed. The conclusions of this study are as follows:

1) The tool diameter and plunge depth are directly proportional to the maximum shear load of the weld. The maximum shear load is obtained by the combination of tool diameter parameters and plunge depth, 2 mm and 300 μm respectively.
2) It was found that the smallest tool diameter and plunge depth was 2 mm and 300 μm respectively, with a maximum shear load of 170.9 N. Then the maximum shear load was 546.4 N in the sample with a successive tool diameter and plunge depth 6 mm and 500 μm. The most optimum joining was obtained in the sample with a tool diameter of 6 mm and a plunge depth of 500 μm where a flat profile with the shortest hook defect was produced.
3) The shape of the fracture that occurs using a 2 mm and 4 mm tool diameter is a shear rupture, while the fracture that occurs with a 6 mm tool diameter with a variation of plunge depth is a plug rupture. It is estimated that the greater the plunge depth, it may minimize the hook defect.

Acknowledgement
This research is supported by the PTUPT Research Grant in 2019 program of Directorate of Research and Public Services, Universitas Indonesia with contract number NKB-1726/UN2.R3.1/HKP.05.00/2019.

Reference
[1]. Mishra, R.S. and Z.Y. Ma, Friction stir welding and processing. Materials Science and Engineering: R: Reports, 2005. 50(1): p. 1-78.
[2]. Thomas W.M., N., E.D., Needham, J.C., Murch, M.G., Templeshmith,P., Dawes, C., Improvements to Friction Welding, in GB Patent Application. 1991.
[3]. Yuan, W., et al., Effect of tool design and process parameters on properties of Al alloy 6016 friction stir spot welds. Journal of Materials Processing Technology, 2011. 211(6): p. 972-977.
[4]. Nee Joo Teh, Helen Goddin, and A. Whitaker, Developments in micro applications of friction stir welding. www.twi-global.com.
[5]. Yin, Y.H., et al., Hook formation and mechanical properties in AZ31 friction stir spot welds. Journal of Materials Processing Technology, 2010. 210(14): p. 2062-2070.
[6]. Wang, D.-A., et al., Mechanical characterization of friction stir spot microwelds. Journal of Materials Processing Technology, 2010. 210(14): p. 1942-1948.
[7]. Baskoro, A.S., et al. Effects of welding parameters in micro friction stir lap welding of aluminum A1100. Trans Tech Publ.
[8]. Baskoro, A.S. and G. Kiswanto. Effects of High Speed Tool Rotation in Micro Friction Stir Spot Welding of Aluminum A1100. Trans Tech Publ.
[9]. Tozaki, Y., Y. Uematsu, and K. Tokaji, A newly developed tool without probe for friction stir spot welding and its performance. Journal of Materials Processing Technology, 2010. 210(6): p. 844-851.
[10]. Aluminium Professionals Australia. "Product data sheet Aluminium 1100 UNS A91100. Available from: http://allyproaustralia.com.au/js/1100-Data%20sheets.pdf.
[11]. Li, W., et al., Improving mechanical properties of pinless friction stir spot welded joints by eliminating hook defect. Materials & Design (1980-2015), 2014. 62: p. 247-254.

IOP Conf. Series: Materials Science and Engineering **727** (2020) 012009 doi:10.1088/1757-899X/727/1/012009

Parametric Analysis and Optimization of Rotary Ultrasonic Machining of Zirconia (ZrO_2) Ceramics

Basem M. A. Abdo[1,2*], Abdulaziz El-Tamimi[2] and Hisham Alkhalefah[1]

[1]Advanced Manufacturing Institute, King Saud University, Riyadh, Saudi Arabia
[2]Industrial Engineering Department, College of Engineering, King Saud University, Riyadh 11421, Saudi Arabia

[*]Corresponding author: babdo@ksu.edu.sa

Abstract. Micro-channels are considered as the essential part of several medical applications such as microfluidics devices, lab on a chip, microbiology, etc. In most applications, fluid flow is required to pass through certain microchannels. The fluid flow characteristics such as flow rate and fluid dynamics are mainly accomplished based on the surface quality and dimensional accuracy of the microchannels. In this study, the microchannels are fabricated on zirconium oxide (ZrO_2) using rotary ultrasonic machining (RUM). A full factorial design of experiments is employed in order to detect the influences of key input parameters of RUM including cutting speed (S), feed rate (FR), depth of cut (DOC), frequency (F) and amplitude (A) on surface roughness (Ra), edge chipping (EC), depth error (DE) and width error (WE) of the milled microchannels. Multi-objective genetic algorithm (MOGA) was employed to determine the optimal parametric conditions for minimizing the values of Ra, EC, DE, and WE of the fabricated microchannels. Results revealed that the minimum value of Ra = 0.26 μm, EC = 10.1 μm, DE = 4.2% and WE = 7.2% can be accomplished through the multi-objective optimization at higher levels of frequency and amplitude and lower levels of feed rate, depth of cut and cutting speed.

1. Introduction

Advanced ceramics such as alumina, zirconia, aluminum nitride, silicon nitride, and boron carbide are extensively being used in various medical and engineering applications. This is due to their superior properties, which include very high hardness, excellent wear resistance, and chemical inertness. However, because of their high hardness and brittleness, the advanced ceramics are difficult to be machined efficiently with traditional machining technologies such as turning, milling, and drilling [1]. Among the advanced ceramics, zirconium dioxide (ZrO_2), commonly known as zirconia, is one of the most widely used ceramic material because of its ability to guarantee excellent properties such as, high hardness, high thermal resistance, and fracture toughness, high flexural strength even at high temperatures and is chemically inert [2]. Due to its high biocompatibility, ZrO_2 is widely used for load-bearing functional structures in medicine and dentistry bio-implantable applications, including joints and dental implants [3]. Micro-channels are considered as the essential part of ZrO_2 for several medical applications such as microfluidics devices, lab on chip, microbiology, etc. In most applications, fluid flow is required to pass through certain microchannels. However, microchannels, micro holes, and other similar micro-features are almost impossible to generate in advanced ceramics through conventional machining processes. Therefore, there is a need to develop a cost-effective nontraditional process to machine advanced ceramic material efficiently.

Several machining processes have been applied for machining zirconia ceramic materials. For example, Guarino et al. [4] used a fiber laser process for milling pockets on yttria-stabilized zirconia. The effect of laser parameters was investigated on surface roughness and material removal rate (MRR). Results showed that the smooth surface along with higher MRR can be achieved at high scan speed, a high number of repetitions and low hatch distance. Nanosecond laser was applied by Jian Li et al. [5] for machining micro-sized steps and a blind hole in yttria-tetragonal zirconia polycrystal ceramic. Results revealed that a high removal rate of 1.35 mm^3/min and smooth surface roughness of Ra = 2.824 µm can be obtained at a scan speed of 150 mm/s and laser fluence of 8.9 J/cm2. Micropatterns of 30 µm diameter and 10 µm depth were fabricated on zirconia by using the Femtosecond laser process in [6]. Laser-assisted grinding was also applied for machining zirconia [7]. Results showed that grinding force the tool damage can be reduced with the help of the laser assistance. The micro electric discharge machining process was applied for precision micromachining of zirconia ceramic. Microchannels with different sizes were fabricated on ZrO_2 by using the laser beam process [8]. The authors investigated the significance of laser beam parameters on the surface roughness and the channels' geometrical accuracy. Results reveal that the surface roughness and the geometrical accuracy can be controlled by adjusting the input parameters of the laser process. The influence of micro-milling parameters was investigated on microgrooves machined on zirconia ceramic in the study documented in [9]. Authors found that the sintering temperature has a major influence on the machinability of ZrO_2. Recently, Rotary Ultrasonic machining (RUM) is applied for machining advanced ceramics. For example, Abdo et al. [10,11] used RUM for machining microchannels on alumina ceramics. Results indicated that RUM can be applied to fabricate very precise microchannels with superior surface finish and high dimensional accuracy under certain optimized RUM input parameters. RUM was also applied for face milling of zirconia ceramic in the study documented in [12]. A mathematical model was developed by Xiao et al. [13] to analyze the cutting force in RUM of dental zirconia ceramics.

From the reported research, it can be concluded that very few research studies are conducted on the micro-machining of zirconia ceramic while no study carried out on the fabrication of microchannels in zirconia by applying RUM. The majority of the reported studies are limited to applying the laser process on the machining of zirconia. The most drawbacks of the laser process are that the process poorly affecting the thermal and the physical the machined surfaces. In this study, a full factorial design is used to analyze the effect of five RUM input parameters on the selected outputs. The selected input parameters include cutting speed (S), feed rate (FR), depth of cut (DOC), amplitude (A), and frequency (F). The outputs including the surface roughness (R_a), edge chipping (EC), channel depth error (DE) and width error (WE) are analyzed. Moreover, the multi-objective genetic algorithm (MOGA) tool is used to optimize the R_a and EC while keeping the DE and WE less than 10%.

2. Materials and method

The experiments in this study were conducted using Ultrasonic 20-linear from DMG, Germany (see figure 1a). Zirconium Oxide (ZrO_2) has been used as a substrate material for micro-machining from CeramTec Germany. All the ZrO_2 samples used in the experiments were of the dimension 50 mm × 50 mm ×10 mm as shown in Figure 1b. Nickel bonded diamond RUM tools with a diameter of 0.5 mm provided by Schott company Germany were utilized in the current investigation as presented in Figure 1c., The cross-sectional dimensions of the fabricated microchannel are 0.5 mm (width) × 0.2 mm (depth) × 6mm (length) (see Figure 2a), The mentioned size of the microchannels was selected based on the micro-fluidics applications [14–16].

2.1 Design of experiments

The full factorial design of experiments (DOE) was used to design the main experiments aiming to detect the influence of RUM input parameters on the selected outputs. Five input parameters with two levels each resulted in 32 experiments (2^5). Eight central points were added randomly to the design, so a total of 40 experiments was performed. The range of the RUM input parameters (see Table 1) used in this study were selected based on the preliminary runs and the studies reported on the micro-machining of glass and ceramic materials using RUM [10,11,17,18].

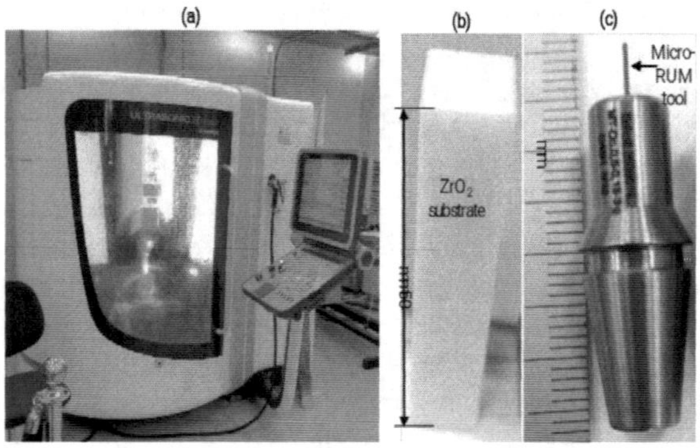

Figure 1: (a) DMG Ultrasonic machine, (b) ZrO_2 substrate, (c) micro-RUM tool

Table 1. RUM parameters and their levels

Parameter (unit) / levels	Lower level	Center point	Higher level
Cutting speed (S), rpm	3000	5000	7000
Feed rate (FR), mm/min	0.5	1	1.5
Depth of cut (DOC), mm	0.0125	0.025	0.0375
Amplitude (A), μm	5	20	35
Frequency (F), kHz	22	24	26

2.2 Measurement Procedures

The channel width, channel depth, were measured using an optical microscopic. The depth error is calculated as the difference between the targeted depth and the measured depth of the channel after machining. Similarly, the width error is calculated as the difference between the targeted width and the actual width after machining. The percentage of width error and depth error were used for the analyzes. Figure 2a represents samples of actual microchannels fabricated on ZrO_2 material and Figure 2b shows a schematic of the microchannels. Surface roughness (Ra) was measured along each channel bed at four different locations using portable surface roughness from Mitutoyo, Japan. The average of these four readings was used for analysis. Figure 2c shows the setup of measuring surface roughness.

3. Results and analysis

Table 2 lists the experimental results of 40 microchannels fabricated in ZrO_2 using full factorial design. Figure 3 shows typical scanned 2D profiles of the microchannels machined at different process parameters. It shows clearly that the RUM input parameters have a significant effect on the surface roughness of the fabricated microchannels.

3.1 Effects of RUM parameters on Ra

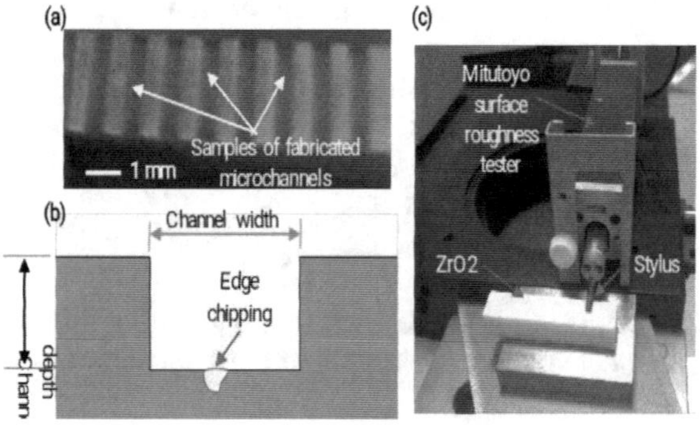

Figure 2. (a, b) Actual and schematic of the machined microchannel, (c) Set up of measuring Ra.

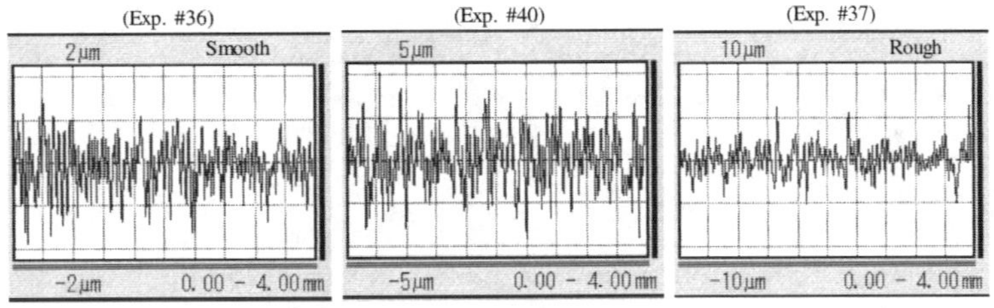

Figure 3. Examples of typical scanned Ra profiles of the fabricated microchannels

Table 2. Experimental results

Exp. #	Input parameters					Outputs			
	S (rpm)	FR (mm/min)	DOC (mm)	F (kHz)	A (µm)	Ra (µm)	EC (µm)	DE (%)	WE (%)
1	7000	0.5	0.0375	26	35	0.42	14.2	5.5	12.1
2	5000	1	0.025	24	20	0.37	15.2	4.6	11.8
3	7000	0.5	0.0375	22	5	0.29	16.8	4	12.4
....	....	....	....	....	....	....	....	....	....
36	5000	1	00.25	24	20	0.23	16.2	3.8	8.7
37	7000	1.5	0.0375	26	35	0.54	14.7	8.1	12.4
38	3000	1.5	0.0375	26	5	0.48	18.7	5.4	12.6
39	7000	0.5	0.0125	26	35	0.26	13.1	7.2	6.3
40	7000	0.5	0.0125	22	35	0.31	9.5	6.9	8.1

The response surface plots for the surface roughness are shown in Figure 4. It is obvious from Figure 4a that the minimum value of Ra can be found at higher levels of cutting speed (S) and lower value of feed rate (FR). This is because the contact time between the machined surface and the RUM tool increases at higher cutting speed, resulting in reducing the cutting force and as a result producing smoother surface [10,19]. On the other hand, the increase in the feed rate causes an increase in Ra. This is because the high feed rate leads to more stress and accordingly produces a rough surface [11,20]. The higher levels of amplitude and moderate to high levels of frequency result in the minimum values of Ra, as depicted in Figure 4b. The surface roughness (Ra) was observed to increase with increasing the depth of cut as shown in Figure 4c.

3.2 Effects of RUM parameters on EC

In RUM, edge chipping (EC) is the main reason for damaging the machined surface [21]. Figure 5 shows the surface plots of the EC. It can be shown in Figure 5a that as the spindle speed increases and the feed rate decreases, the values of EC reduce considerably. This happens because any change in input parameters which leads to an increase in the force means the edge chipping will increase [10,11]. Regarding the effect of ultrasonic parameters, it can be said from Figure 5b that the minimum values of edge chipping are found at high levels of frequencies and amplitudes. Due to an increase in the cutting force, the EC was observed to increase with increasing the depth of cut as shown in Figure 5c.

3.3 Optimization results

A multi-objective genetic algorithm (MOGA) was applied to minimize Ra and EC of the fabricated microchannels. The constraints were to keep the depth error and the width error of less than 10%. A total of 774 generations were modeled using the MOGA. Figure 6 shows the design points obtained using the bubble charts. The real design points represented to the actual experimental runs and the virtual runs are predicted by the response surfaces based on radial basis functions.

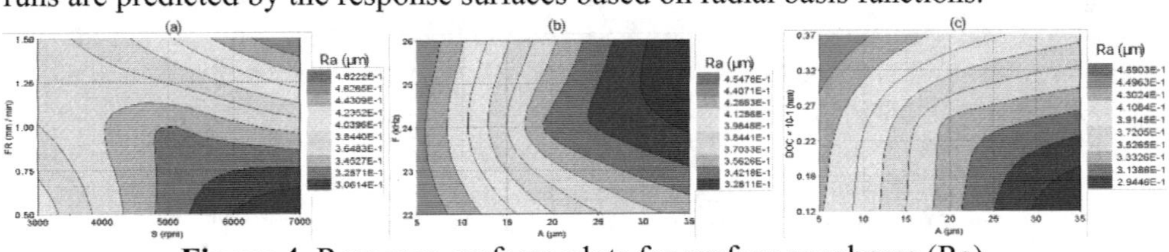

Figure 4. Response surfaces plots for surface roughness (Ra).

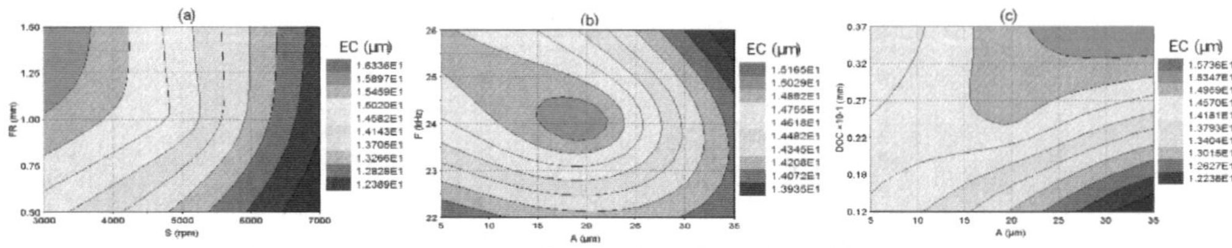

Figure 5. Response surfaces plots for edge chipping (EC).

The four outputs at a time can be shown in Figure 6 using a 4D bubble. The colors of the bubbles represent the width error (WE) whereas the diameter represents the depth error (DE). The effect of Ra and EC is clearly recognized in the 4D graph, as the low Ra corresponds to low to med EC. These design points are described by high cutting speed, amplitude and frequency, and low feed rate and depth of cut. On the other side, the above right area of Figure 6 corresponds to design points with high Ra and EC. These are related to high feed rate and depth of cut, low cutting speed, amplitude, and frequency. Regarding the WE and the DE, it can be noticed from the difference in the colors and size of the bubbles in Figure 6 that the variation is not high, and most of the values are located between 5 to 10 µm.

The relationships between all input and output parameters can be seen using a parallel coordinate chart as shown in Figure 7. Most of the unfeasible design points are linked to higher values of feed rate (FR), depth of cut (DOC) and cutting speed which leading to higher width error (WE) and depth error (DE). No significant effect of Frequency (F) and amplitude (A) on WE and DE can be found in Figure 7 as the unfeasible points distribute overall levels of both F and A. Looking to overall responses, the optimal results were found at low levels of feed rate, cutting speed and depth of cut and high levels of amplitude and frequency as listed in Table 3.

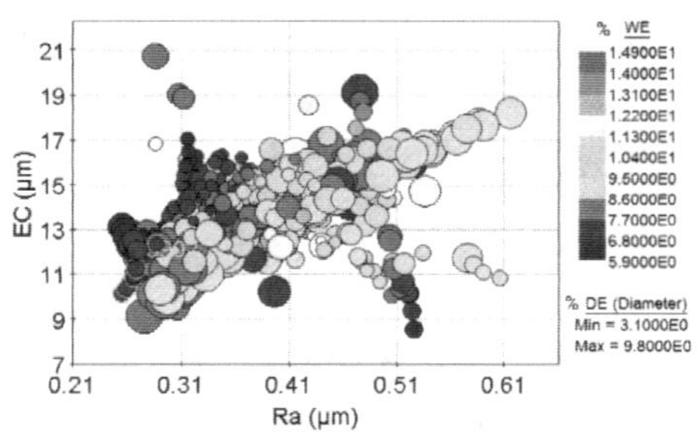

Figure 6. A 4D bubble chart showing the design points obtained with outputs of Ra, EC, WE, and DE

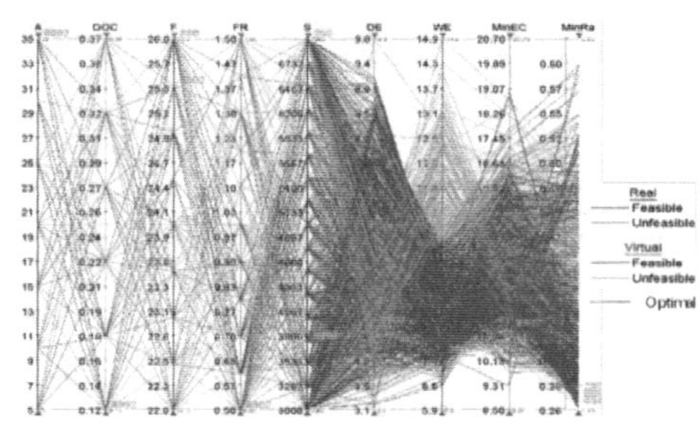

Figure 7. A parallel coordinate chart for the analysis of RUM parameters.

Table 3. Optimal points for all outputs

#	S (rpm)	FR (mm/min)	DOC (mm)	F (kHz)	A (µm)	Ra (µm)	EC (µm)	DE (%)	WE (%)
1	3000	0.5	0.0125	26	35	0.26	10.1	4.5	7.2
2	3500	0.5	0.0125	26	35	0.27	10.64	4.70	7.23
3	4000	0.6	0.0125	26	35	0.27	11.28	4.67	7.33
4	3250	0.5	0.0125	26	35	0.26	10.38	4.60	7.21

4. Conclusions

- RUM can be applied to fabricate precise microchannels in ZrO_2 material. Under the experimental condition, the dimensional accuracy in terms of the depth error and width error is found to be less than 10% and 13% respectively.
- Ra is directly proportional to the depth of cut and feed rate and inversely with cutting speed, frequency (F) and amplitude (A).
- Edge chipping (EC) reduces with increasing cutting speed and decreasing the feed rate and depth of cut.

- Increasing of ultrasonic parameters (F and A) leading to increasing EC.
- The optimal RUM parameters are a feed rate of 0.5 mm/min, a depth of cut of 0.0125 mm, a frequency of 26 kHz, a cutting speed of 3000 rpm, and an amplitude of 35 μm, resulting the outputs as Ra = 0.26 μm, EC = 10.1 μm, DE = 4.2%, and WE = 7.2%.

References

[1] Rayat M S, Gill S S, Singh R and Sharma L 2017 Fabrication and machining of ceramic composites — A review on current scenario *Mater. Manuf. Process.* **32** 1451–74
[2] Daou E E 2014 The Zirconia Ceramic: Strengths and Weaknesses *Open Dent. J.* **8** 33–42
[3] Abd El-Ghany O S and Sherief A H 2016 Zirconia based ceramics, some clinical and biological aspects: Review *Futur. Dent. J.* **2** 55–64
[4] Guarino S, Ponticelli G S, Giannini O, Genna S and Trovalusci F 2017 Laser milling of yttria-stabilized zirconia by using a Q-switched Yb : YAG fiber laser : experimental analysis
[5] Li J, Ji L, Hu Y and Bao Y 2016 Precise micromachining of yttria-tetragonal zirconia polycrystal ceramic using 532 nm nanosecond laser *Ceram. Int.* **42** 4377–85
[6] Stanciuc A M, Flamant Q, Sprecher C M, Alini M, Anglada M and Peroglio M 2018 Femtosecond laser multi-patterning of zirconia for screening of cell-surface interactions *J. Eur. Ceram. Soc.* **38** 939–48
[7] Kizaki T, Ito Y, Tanabe S, Kim Y, Sugita N and Mitsuishi M 2016 Laser-assisted Machining of Zirconia Ceramics using a Diamond Bur *Procedia CIRP* **42** 497–502
[8] Abdo B M A, Ahmed N, El-Tamimi A M, Anwar S, Alkhalefah H and Nasr E A 2019 Laser beam machining of zirconia ceramic : An investigation of micro-machining geometry and surface roughness † **33** 1817–31
[9] Twardy S, Stolle J, Riemer O, Treccani L, Brinksmeier E and Rezwan K 2012 Micromachining of ceramic surfaces : Hydroxyapatite and zirconia *J. Mater. Process. Tech.* **212** 614–24
[10] Abdo B M A, Anwar S, El-Tamimi A M and Nasr E A 2019 Experimental Analysis on the Influence and Optimization of μ -RUM Parameters in Machining Alumina Bioceramic
[11] Abdo B M A, Anwar S and El-Tamimi A 2019 Machinability study of biolox forte ceramic by milling microchannels using rotary ultrasonic machining *J. Manuf. Process.* **43** 175–91
[12] Abdo B M A, Darwish S M and El-Tamimi A M 2012 Parameters optimization of rotary ultrasonic machining of zirconia ceramic for surface roughness using statistical Taguchi's experimental design *Appl. Mech. Mater.* **184–185**
[13] Xiao X, Zheng K and Liao W 2014 Theoretical model for cutting force in rotary ultrasonic milling of dental zirconia ceramics *Int. J. Adv. Manuf. Technol.* **75** 1263–77
[14] Whitesides G M 2006 The origins and the future of microfluidics *Nature* **442** 368–73
[15] Farhan Shafique M, Laister A, Clark M, Miles R E and Robertson I D 2011 Fabrication of embedded microfluidic channels in low temperature co-fired ceramic technology using laser machining and progressive lamination *J. Eur. Ceram. Soc.* **31** 2199–204
[16] Ong S E, Zhang S, Du H J and Fu Y Q 2008 Fundamental principles and applications of microfluidic systems *Front. Biosci.* **13** 2757–73
[17] Wang J, Shimada K, Mizutani M and Kuriyagawa T 2018 Effects of abrasive material and particle shape on machining performance in micro ultrasonic machining *Precis. Eng.* **51** 373–87
[18] Sandeep K and Akshay D 2018 Fabrication of microchannels using rotary tool micro-USM: An experimental investigation on tool wear reduction and form accuracy improvement *J. Manuf. Process.* **32** 802–15
[19] Xiao X, Zheng K, Liao W and Meng H 2015 Study on cutting force model in ultrasonic vibration-assisted side grinding of zirconia ceramics *Int. J. Mach. Tools Manuf.* **104** 58–67
[20] Singh R P and Singhal S 2016 Investigation of Machining Characteristics in Rotary Ultrasonic Machining of Alumina Ceramic *Mater. Manuf. Process.* **32** 309–26
[21] Wang J, Zhang J, Feng P and Guo P 2018 Damage formation and suppression in rotary ultrasonic machining of hard and brittle materials : A critical review *Ceram. Int.* **44** 1227–39

IOP Conf. Series: Materials Science and Engineering **727** (2020) 012010 doi:10.1088/1757-899X/727/1/012010

A Study on the Change of Manufacturing Design Process due to the Development of A.I Design and 3D Printing

Chun Hyunjin[1,*]

[1] School of Arts, Nanjing University of Aeronautics and Astronautics, No. 29 JIangjung Road, Jiangning District, Nanjing 211106, Jiangsu, China

[*]Corresponding E-mail: tough4324@naver.com

Abstract. The fourth industrial revolution brought about changes in various fields. The speed of the fourth industrial revolution will proceed faster than the previous ones. It is a new revolution in which new technologies are integrated. So the convergence of new technologies in the fourth industrial revolution will affect many areas. In particular, new technologies such as artificial intelligence and 3D printing technology are being used in the manufacturing industry. The fourth industrial revolution will change the manufacturing design process in the future. Predicted changes are as follows: First, existing management processes are divided into planning, design and manufacturing processes. The collection and analysis of many data are important in the planning process. This planning phase enabled data collection, processing and management through cloud computing and big data technologies. Based on these plans, artificial intelligence can produce thousands of design results. Next, based on the design plan, products can be produced in 3D printing. As such, the manufacturing process has been very simplified by 3D printing. The study can offer some solutions for social changing problems by predicting the future of manufacturing. And the results of the study can be used as a research base in the manufacturing design sector using artificial intelligence and 3D printing.

1. Introduction

1.1. Research Background and Purpose

Recently, the Fourth Industrial Revolution has brought about changes in various fields. The fourth industrial revolution is a convergence of information intelligence technologies such as artificial intelligence, Internet and big data, as well as new technologies such as 3D printing and robots. So all the products are connected and the product can solve problems on its own. The speed of the Fourth Industrial Revolution will proceed faster than the First, Second and Third Industrial Revolution. Some scholars argued that the Fourth Industrial Revolution was an extension of the Third Industrial Revolution. But the term "fourth industrial revolution" was first used at the World Economic Forum in 2016. The World Economic Forum declared that the Fourth Industrial Revolution would be different from the Third Industrial Revolution. The Fourth Industrial Revolution is a new revolution in which new technologies are integrated. So the convergence of new technologies in the Fourth Industrial Revolution will affect various areas. In particular, new technologies such as artificial intelligence and 3D printing technology are being used in the manufacturing field. While this is not yet practical due to the lack of technology in manufacturing, a variety of new technologies will be used in manufacturing in the future. It is expected that this technology will bring significant changes to the manufacturing design process as well. The environment of the design industry such as the fashion industry is rapidly changed from the human-involved production process to the IT and AI-based automatic production process (Jung Ju Ri etc. 2019). In particular, the manufacturing sector has a lot of interest at the government level. So a variety of researchers are researching manufacturing. However, while there is a lot of research on technology and methods in manufacturing design, there is a lack of research on the changes and social implications that these technologies will bring to the design industry. So in this study, current technologies for artificial intelligence and 3D printing were analyzed. Then, based on the results of the analysis, the applicability was analyzed in the manufacturing design field. It also predicted changes in society. This study can prepare for social change by predicting the future of the manufacturing industry. And the results of this study can be used as a basis for research in the field of manufacturing design using artificial intelligence and 3D printing.

1.2. Literature Review

The Fourth Industrial Revolution is recognized by researchers as an important issue. Thus, advanced countries are conducting various researches, including artificial intelligence related to the Fourth Industrial Revolution. So this study summarized the researches that were done concerning artificial intelligence and 3D printing. Lee Jae Ik (2018) suggested the direction of the next-generation convergence design based on the important issues and alternatives in each area, by analyzing the characteristics of the 4th industrial revolution and the design convergence of the 4th industrial revolution era [1]. Jung Ju Ri etc. (2019) examined the change in the environment of fashion design resulting from the fourth industrial revolution among AI-based innovative cases, and further draw its characteristics [2]. Kim Hye Eun (2015) investigated the changes taking place in the industrial field through the development of 3D printing and its meanings, and forecasting the changes in fashion and society through its popularisation and generalization, exemplified by a case study of 3D printing in fashion [3]. Jeong Won Joon etc. (2018) proposed the role of future designers and capabilities to be developed in the age of A.I [4]. Han Jung yeob (2013) tried a comprehensive study on applying domestic and foreign 3D printing case to the analysis sheet [5]. Chun Ha bong (2017) analyzed on the convergence service design for designing users' experience, through which users make conclusive data at the stage of using smart-connected product and service in daily life, extracted a concept and structure model of convergence design, through phases of business model, combination and analysis at the stage of use, and finally researched methodical access for corporate's strategic utilization, for the purpose of designing the business model at such stage of use [6]. In previous studies, many studies have been conducted on the technical methods and solutions for manufacturing related to the Fourth Industrial Revolution. However, there is still a lack of research on how manufacturing industries can change due to the development of new technologies. Besides, the fourth industrial revolution is to converge information technology such as big data and mobile with new technologies such as artificial intelligence and robotics. There is also a lack of research on this convergence of intelligence information technology and new technologies. So in this study, I conducted a study on the convergence of artificial intelligence and 3D printing. Also, the application of these technologies in manufacturing and design was analyzed in this study. Lastly, the changes in the manufacturing design process were analyzed through these applications. These studies may be the theoretical basis for the field of manufacturing design using new technology.

2. Research Method

The research method in this study analyzed the characteristics and processes of manufacturing design through literature research related to artificial intelligence and 3D printing. In the literature study, the technology of artificial intelligence and 3D printing was analyzed through related papers, reports, and online data. It also analyzed the current situation of the manufacturing industry and predicted the future manufacturing design process due to the Fourth Industrial Revolution. In this study, I compared the existing manufacturing design process with the future manufacturing design method. The main research scope of this study is the design area. Through these changes in the design area, the future of

IOP Conf. Series: Materials Science and Engineering **727** (2020) 012010 doi:10.1088/1757-899X/727/1/012010

the manufacturing industry was analyzed.

3. A Theoretical Study on A.I Design

3.1. Artificial Intelligence

Artificial intelligence is a technology that a computer performs on its own, such as human learning, awareness, and understanding. In particular, the artificial intelligence field has developed as the development of the Internet has led to the collection of a lot of data. This is how computers collect data and learn on their own. A machine is a way to analyze certain patterns based on these data, predict future outcomes, and make appropriate decisions. This principle is based on the theory of neural net. A network is formed by connecting synapses to a machine as if it had nerves in humans. The machine can solve problems by changing the connection of these synapses. Artificial intelligence technology will enhance human life quality. And these artificial intelligence technologies will be utilized in various fields [4].

3.2. Artificial Intelligence Design

As artificial intelligence technology developing, it is being used in various fields. In particular, artificial intelligence is being used in the field of design that is in desperate need of creativity. Currently, a generative design technology based on artificial intelligence has been developed. Generative designs are provided with numerous design results when designers put into materials, manufacturing methods and prices. These technologies are based on artificial intelligence's algorithms and cloud computing technology [4], which can produce hundreds of design results within a very short time. The process of creative design is as follows (figure 1). [7].

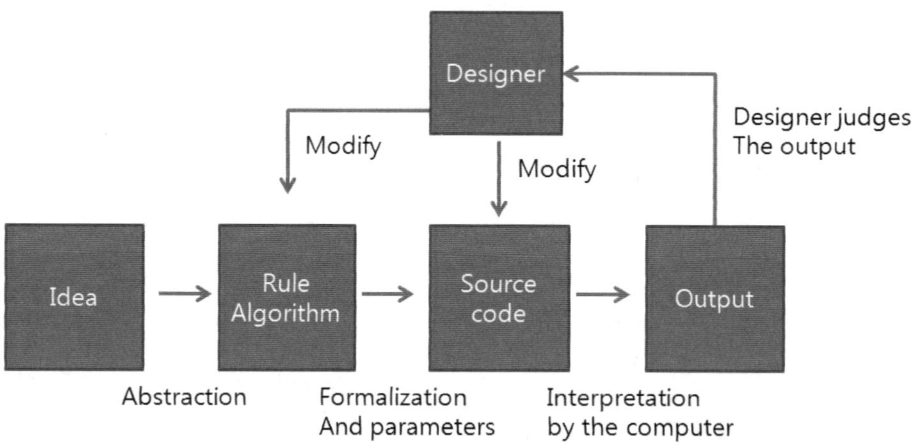

Figure 1 Process of generative design [7]

Artificial intelligence technology will shrift the design realm where creativity is required and necessary through 'clouding computing' technology that solves computation and 'dip running' that accumulates knowledge on its own. The generative design presents a reasonable design considering structural engineering, economics, and aesthetics when the designer enters the price and material into the computer after performing basic sketches. At the University of California, Berkeley, artificial intelligence's Philip Isola research team is mainly conducting creative design research. By sketching an approximate product and typing it into a computer, the team can create thousands of designs through artificial intelligence [7-8]. These results can replace major tasks of designers by artificial intelligence. The results produced by the Philip Isola research team are as shown in Figure 2 [8]. These studies will be used in architectural design that is more complex than product design. So far, artificial intelligence has not produced creative results compared to designers. However, in the future, AI will create creative designs through digital platforms.

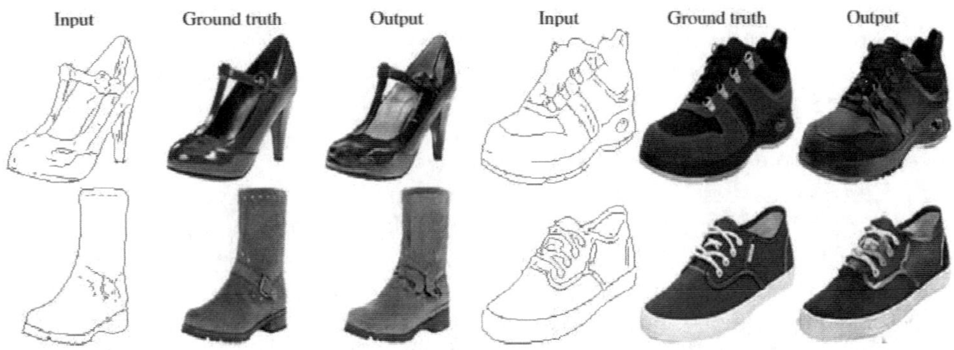

Figure 2 Example of product design using a creative design [8]

4. A Theoretical Study on 3D Printing

4.1. 3D Printing

3D printing is a method that completes product form by stacking digital files one layer at a time with the Additive Manufacturing method. 3D printing can complete a model using a three-dimensional modeling file using a rigid material. It has no special processes, allowing producers to make products at lower prices while bearing the advantage of being able to create necessary objects without limitations of time and space. 3D printing, as it were, is expected to revolutionize manufacturing. With the launch of the "Maker Movement" worldwide, 3D printing technology is becoming popular.

4.2. Application of 3D Printing

As 3D printing technology emerging, businesses that utilize 3D printing are becoming more popular around the world. For example, a private operator uses 3D printing to produce and sell small products. Hospitals also adopt 3D printing technology to make artificial organs. Furthermore, a company has developed a technology to make food through 3D printing, by which various types of food can be made using 3D printing. Because of these 3D printing technologies, drawing models for 3D printing is also becoming important. Many businesses use drawings for 3D printing. Producers who are not familiar with design may purchase 3D printed drawings created by designers. A site called 'Thingiverse', for example, offers a service to share drawings for 3D printing. The service is a way to make money by selling 3D printed drawings made by designers. This development of 3D printing has democratized the manufacturing industries. Capital led manufacturing after the Second Industrial Revolution. However, the development of 3D printing has special and practical meaning that even the general public can do manufacturing.

5. Change of Manufacturing Design Process

5.1. Current Process of Manufacture

The current manufacturing process is generally divided into three main categories. These three categories are planning, design and manufacturing. And the planning stage is further divided into research, design requirements, feasibility assessment, and establishing design requirements. The design stage includes concept design, primary design, and specified design. And finally, it consists of the manufacturing stage. At the planning stage, it is to analyze the market's needs and problems. Also, this step analyzes the validity of the product and the requirements of the design, which determines whether a product can be made or not. These investigations are conducted by experienced engineers. In the design stage, ideas are generated and implemented based on the data analyzed in the planning stage. In particular, the primary design is the stage in which the whole system is created. Therefore, the primary design is a step that connects the concept of design with a specific design. And the specific design is to create a complete design through modeling and drawing. At this stage, designers can design efficiently using the Computer-Aided Design program. The manufacturing stage plans and determines how products are produced. The engineer then selects the production process and performs mass production. At this time, prototype tests are conducted to determine if they are suitable for mass production.

5.2. Future Prediction Analysis of Manufacturing Design Process

The future manufacturing design process will be changed by the Fourth Industrial Revolution. Predicting these changes are as follows (figure 3): First, the existing management process is divided into the Planning, Design and Manufacturing stage. In the planning process, the collection and analysis of many data are important. These parts can be solved a lot through artificial intelligence. In particular, the planning stage has enabled data collection, processing and management through cloud computing and big data technologies. This allows designers to process large amounts of information quickly [4]. Based on these plans, the approximate design direction is determined, and designers can input elements of design, such as goals, prices, and materials, into the algorithms of the computer, and artificial intelligence can produce thousands of design results. So far, designers have completed one design through the final analysis, but in the future, artificial intelligence can complete thousands of designs through analysis [4]. And the completion of this design allows consumers to easily change color or shape through computer programs. Next, after AI completes the design, it can manufacture it through 3D printing. So far, manufacturing has been mass-producing by lowering prices. But the development of 3D printing will change the way it is produced. 3D printing is being able to produce small amounts and individual products. So there will appear many small companies in the manufacturing sector. As a result, consumers will be more satisfied with their products. There will also be a major shift in manufacturing as individuals' 3D printing becomes common [2]. As such, the manufacturing process has been very simplified by 3D printing [2].

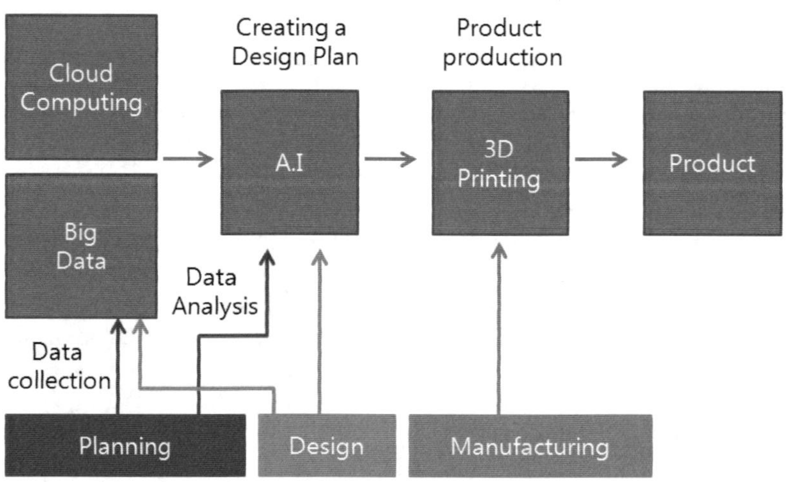

Figure 3 Future Manufacturing Design Process

6. Change in the Manufacturing Industry

The invention of artificial intelligence and 3D printing will make a big difference in the manufacturing industry. Especially when artificial intelligence technology develops, computers will replace many of the designers' work. And it takes the collection and analysis of a large number of data for manufacturers to design and produce products. These parts will be logically analyzed by artificial intelligence through big data. So in the future, artificial intelligence will provide the most appropriate alternative for design through the analysis of data. For this reason, much in the field of product planning and design will be replaced by machines. But innovation originates from human imagination and it is difficult to replace with technology. Even if artificial intelligence develops, it is not easy to replace human creativity. So creativity is the most important value for designers in the future [4]. Therefore, designers who do not have well creative abilities are doomed to lose jobs. Given this, in product design education, improve creativity is especially crucial. And 3D printing will revolutionize manufacturing based on diversity and efficiency. Manufacturing will change from mass production to small production. So products will be made according to consumers' tastes. Besides, consumers will be directly involved in the production. As a result, the Fourth Industrial Revolution will democratize manufacturing. However, many problems arise with these changes in the times. For example, the gap between rich and poor will widen with technology holdings. And problems such as quality control will arise as a single-person manufacturer appears. So the government must come up with institutional devices to solve these problems.

7. Conclusion

The recent Fourth Industrial Revolution will bring about transformation in manufacturing. The Fourth Industrial Revolution is a new revolution in which several technologies are integrated. So the convergence of new technologies in the fourth industrial revolution will affect many areas. New technologies such as artificial intelligence and 3D printing technology are being employed in the manufacturing industry. It is anticipated that new technologies will bring significant changes to the manufacturing design process. As a result, the fourth industrial revolution will produce the democratization of manufacturing. Nevertheless, many social problems arise from these changes in the times. For example, the gap between rich and poor will widen to technology holdings. And with the emergence of a single-person manufacturer, problems such as quality control will spring up. The government must come up with an institutional mechanism to address these problems. In general, the study analyzed the technologies of artificial intelligence and 3D printing about manufacturing and predicted changes in society. It can provide, to some extent, with countermeasures for changes in society by predicting the future of manufacturing. The results of the study can be regarded as the basis for research in the manufacturing design field from the perspective of artificial intelligence and 3D printing.

References

[1] Jung Ju ri etc. 2019 *Study on AI-based Design Process Innovation in Fashion Industry.* Journal of communication design. **0** Vol 67 pp 140-153

[2] Kim Hye Eun 2015 *The research into the changes of fashion industry according to the development of 3D printing technology.* Journal of the Korean Society of Fashion Design. **4** Vol 15 pp 17-33

[3] Lee Jae Ik 2018 *A Study on the Design Convergence and the 4th Industrial Revolution.* Design Research. **3** Vol 3 pp 147-155

[4] Jeong Won Joon etc. 2018 *A Study on the Role of Designer in the 4th Industrial Revolution -Focusing on Design Process and A.I based Design Software-* Journal of Digital Convergence. **8** Vol 16 pp 279-285

[5] Han Jung yeob 2013 *A study on the Prototype Modeling Method using 3D Printing.* Research Bulletin of Package Culture Design. **34** Vol 34 pp 97-109

[6] Chun Ha bong 2017 *A Study on changes in manufacturing-based industries, and types of changes in design areas and its features based on the Fourth Industrial Revolution - Focused on Intelligence-based Convergence Design Strategy-.* A Treatise on the Plastic Media. **20** Vol 1 pp 218-229

[7] H. Bohnacker etc. 2009 *Generative Gestaltung.* Mainz : Verlag hermann schmidt.

[8] Phillip Isola etc. 2018 *Image-to-Image Translation with Conditional Adversarial Networks.* Berkeley AI Research (BAIR) Laboratory, UC Berkeley. **26** arXiv:1611.07004v3 [cs.CV]

IOP Conf. Series: Materials Science and Engineering **727** (2020) 012011 doi:10.1088/1757-899X/727/1/012011

A Study on Application of Generative Design System in Manufacturing Process

Chun Hyunjin[1, *]

[1] School of Arts, Nanjing University of Aeronautics and Astronautics, No. 29 JIangjung Road, Jiangning District, Nanjing 211106, Jiangsu, China

[*]Corresponding E-mail: tough4324@naver.com

Abstract. The paradigm of manufacturing is changing with the recent era of the fourth industrial revolution. In particular, with the rapid development of artificial intelligence technology, tremendous transformation is welcomed in various areas related to manufacturing. This study concentrated on the changes in manufacturing due to the development of a generative design system. It also analyzed how designers design products through a generative design system based on artificial intelligence. The changes in the design part of the manufacturing process through the creative design are as follows: First, once the designer sets up a rough conceptual design, and then enters the conceptual design into the generative design system, the computer provides with a basic form of design. After AI presents multiple designs, designers choose the appropriate and desirable one. Next artificial intelligence puts forward an optimal material proposal considering its function and economics. And the designer reviews this alternative. After the designer's review, the computer comes up with a final proposal. Much of the design is done by artificial intelligence. This design method will be universal in the future. Therefore, this study analyzed the direction in which the generative design system would develop in the artificial intelligence era. And the results of this study are to present the theoretical basis of the future product design methodology.

1. Introduction

1.1. Research Background and Purpose

Due to the recent development of artificial intelligence technology, these technologies are being used in various fields. Artificial intelligence technology has had a lot of impact on the manufacturing industry as it fuses with digital technologies. The manufacturing industry is an area where the characteristics of the times are well-reflected. Manufacturing has grown through the Second and Third Industrial Revolution, reflecting the social phenomena of the times. The paradigm of manufacturing is also changing with the recent era of the fourth industrial revolution. This rapid development of technology is expected to lead to significant changes in various areas within the manufacturing industry. In particular, the introduction of the general design using artificial intelligence is anticipated to bring about many changes for product designers. Therefore, this study attempts to predict the direction in which the general design will be developed in the artificial intelligence era.

1.2. Literature Review

As artificial intelligence technology developing, it is also being applied in various manufacturing industries. These studies have been of great interest to various research institutes, including universities. Studies using A.I. and General Design technologies were as follows. Park Su Bin (2017) studied which kind of art had already employed A.I technology, and how the definition of art and teaching art would be changed in the age of artificial intelligence. Then she analyzed 'Google A.I. Experiments' can be used for an integrated curriculum of art education tools [1]. Lee Jin Wook (2011) investigated application methods and characteristics of fractal geometry and Voronoi diagram which were the most representative principals of generative design in nature by researching the example of furniture design using these principals [2]. Ahn Seong Mo etc. (2009) verified the applicability of eco-system to digital design generation system with a focus on the ecological approach of digital design [3]. Jeon Young Jun (2018) prepared to help understand general social trends through a case study of the past industrial revolution and analyzed the difference between the Fourth Industrial Revolution to assess current trends and technical achievements [4]. Lee Jae Young (2018) proved the efficiency of the new design. It will have a beneficial effect on the creation of aesthetic buildings. In previous studies, the generative design system was recognized as an innovative technology in the field of design. However, there is a lack of research into the relevance between the generative design system and the manufacturing sector. So this thesis investigates the application of the generative design system in the manufacturing process. These studies can be used as the theoretical basis for the manufacturing sector using artificial intelligence in the era of the Fourth Industrial Revolution.

2. Research Method

In this study, a creative design system among artificial intelligence technologies is the main research scope. Generative design systems are implemented through informational parameters and creation algorithm. In the future, artificial intelligence will help designers design their desired outcomes through this technology. Such artificial intelligence will serve to create product forms and perform human creativity, rather than simply as a tool to assist design. Recently, a creative design system based on artificial intelligence is being used in various fields of design. This thesis, in light of this, probed into the changes in manufacturing on account of the generative design. In the meanwhile, this study analyzed the application of product design through creative design. It is also intended to provide a direction for the development of generative design in manufacturing.

3. A Theoretical Study

3.1. Artificial Intelligence

Artificial intelligence is the process of solving problems through computers. Through the development of modern technology, computers go beyond solving simple problems. They, instead, take advantage of their intelligence to work out new solutions. Therefore, artificial intelligence performs learning, understanding, and reasoning through computer programs. In general, humans recognize information through the eyes and understand it through the brain. As such, information obtained through the eyes becomes knowledge. Applying this learning method to computers is called deep learning. And the way computers change information into knowledge is called algorithms [4]. The principles of artificial intelligence are as follows. First, artificial intelligence observes the environment. Through these observations, it not only collects data but also interprets data. Then it determines the reaction of the situation. Then the smart technology observes the results of the action and analyzes the data again. And it determines the next reaction. Artificial intelligence learns on its own by repeating this process. This is the basic process of artificial intelligence learning (figure 1) [4].

IOP Conf. Series: Materials Science and Engineering **727** (2020) 012011 doi:10.1088/1757-899X/727/1/012011

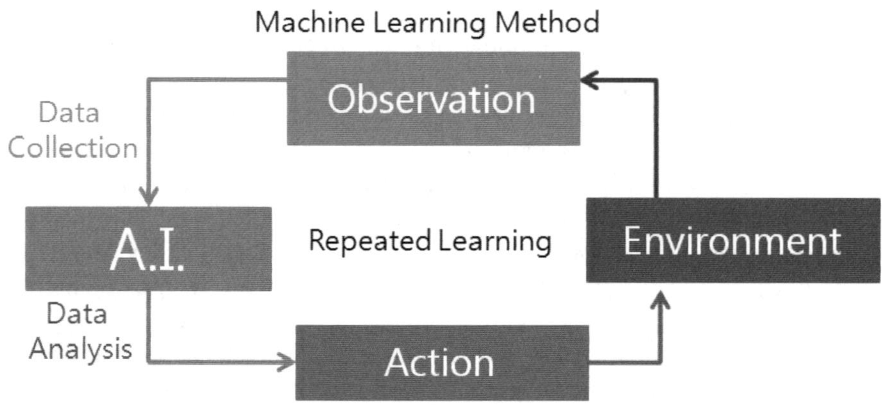

Figure 1 Artificial Intelligence Operation Principle [4]

3.2. Artificial Intellilgence Application

Artificial intelligence technology is being used in various fields. Such artificial intelligence technologies are being conducted around global conglomerates like Google and IBM. Google's artificial intelligence beat world-class Go player Lee Se-dol in Go [1]. And IBM has developed artificial intelligence 'Watson' and is being adopted in the medical field. Watson is an artificial intelligence that accurately analyzes the patient's condition and offers the best treatment. Artificial intelligence technology will improve productivity in a variety of areas. Most of them will be replaced by artificial intelligence, especially in areas where they engage in physical activity. But even in creative fields such as art and design, there are various opinions as to whether artificial intelligence can be a replacement. Recently, there has been a lot of research in the field of art using artificial intelligence. Microsoft, for example, reproduced a picture of Rembrandt with artificial intelligence. Microsoft digitized 346 Rembrandt's paintings and trained them to deep-learning artificial intelligence using digital information. After AI finished learning about Rembrandt's painting style, it ordered artificial intelligence to "draw a white male in his 30s and 40s," which completed a painting that seemed to be drawn by Rembrandt. As such, artificial intelligence is expected to be used in various fields of art and design in the future.

3.3. Generative Design System

A generative design system is due to the development of an artificial intelligence computer system that is actively involved in the design process. A generative design system's core technology is artificial intelligence. The generative design system has no difference from the artist's teaching methods. Through repeated learning by artificial intelligence, artistic works can be completed very quickly. So recently, various software companies are trying to develop a creative design program. A representative company of them is Autodesk, who has developed a program called Dreamcatcher that combines computer-aided design with artificial intelligence (figure 2). When a designer enters basic forms and variables together, artificial intelligence produces a product that considers various factors such as structural engineering and economic efficiency. The result of this design is a product that takes into account various factors such as economy, structure, and beauty. Also, the results created by artificial intelligence can lead to a special form that humans do not think of. If artificial intelligence technology further develops and popularizes in the future, a significant portion of the creative field will be replaced by artificial intelligence. In particular, a generative design system will bring about great variation in the manufacturing process.

IOP Conf. Series: Materials Science and Engineering **727** (2020) 012011 doi:10.1088/1757-899X/727/1/012011

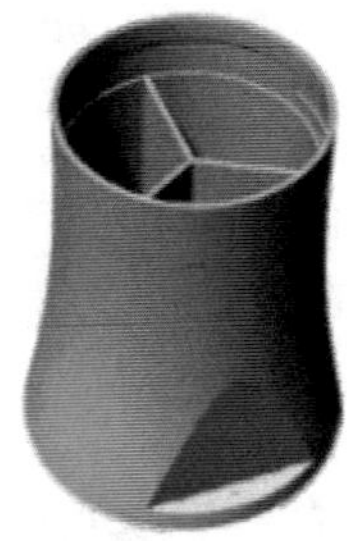

Figure 2 Manufacturing Design Using Dreamcatcher [5]

4. Application of Generative Design System in Manufacturing Process

4.1. The Future of the Generative Design System

Generally, in the manufacturing process, designers needed a lot of experience and knowledge to design their products. Designers spent a lot of time collecting data on products, analyzing materials, and performing structural engineering stability, aesthetics, etc. However, if a general design system using artificial intelligence is established, anyone who understands operate the program can make the product. And the General Design System will work with designers to produce faster and better results, and it will also be able to make optimized products that take into account various constraints, appearance, and materials. After the designer enters a rough sketch into the configured computer system, artificial intelligence makes an overall consideration of various factors and produces the best results. The process of the generative design system is as follows (figure 3) [6, 7].

Figure 3 Generative Design System [6, 7]

4.2. Manufacturing Process Using Generative Design System

The product manufacturing process consists of Propose, Plan, Design, Development, Evaluate, and Production. The manufacturing process is as follows (figure 4).

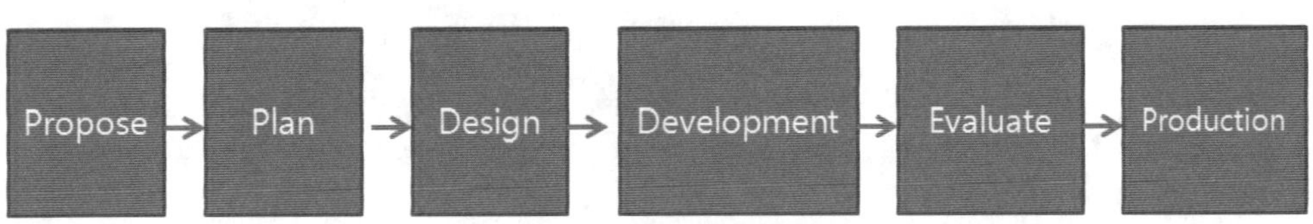

Figure 4 Manufacturing Process

First, Propose is the process of determining the requirements of a project to develop a product. The following process is the Plan. Planning is the stage in which various data are investigated and analyzed to prepare for design. Based on the plan, the designer conducts product design. The following process is Design. Design is divided into conceptual design, basic design, detailed design, and production design. The design phase refers to all four of these steps. This process is a step that embodies the structure and shape of the product. The next step is to Develop, which embodies the design. At this stage, 3D programs such as sketch-ups and 3D MAX are used to refine them. The following process is to Evaluate. At this stage, the product is being modeled and evaluated. And it's a step in determining productivity while testing a product. The final step is Production. This step is the production and distribution of products on a large scale at the factory.

Figure 5 Manufacturing Process Using Generative Design System

There will be a big change in the application of the Generative Design System in the manufacturing process (figure 5). As these digital design technologies advance, a new approach will emerge in the design of the manufacturing sector. The changes in design are as follows: First, the designer sets up an approximate conceptual design. Then, when the designer enters the conceptual design into the Generative Design System, the computer provides the basic design. Designers choose the appropriate design from the basic designs presented by the Generative Design System. Based on the selected design plan, artificial intelligence then presents an optimal material proposal considering its function and economics. And the designer reviews this alternative. After the designer's review, the computer comes up with a final proposal. Most of these designs are performed by artificial intelligence. This design is called a hybrid design (figure 6) [8]. It is a design process that will take place in the future. Designers and artificial intelligence work closely together to produce results. These designs will be utilized not only in product design but also in various areas of design. And a digital platform will be formed through the Internet. So when various data are built, the design will be formed under the influence of the algorithm. Through deep learning, artificial intelligence generates huge data. Computers can review more information than humans. Much of the data accumulated in these programs will be utilized through artificial intelligence and designers will be able to make new forms of design [8].

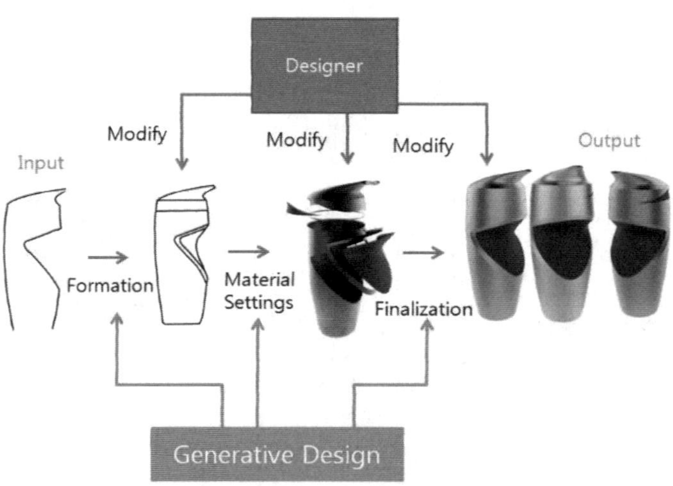

Figure 6 Hybrid Design System

IOP Conf. Series: Materials Science and Engineering **727** (2020) 012011 doi:10.1088/1757-899X/727/1/012011

4.3. Development of Manufacturing Industries through a Generative Design System

With the development of the General Design System, artificial intelligence can create designs that humans never thought of. A new design can be created by calculating the curvature, the variable that has been modified in the design. With the help of artificial intelligence, the parts that are difficult and time-consuming for humans to express will be able to be solved. Besides, the Generative Design System will push beyond analog limits to break down the boundaries of expression and allow virtual space to express real space [9]. As such, in the future, various design attempts will be possible in digital space beyond the physical limitations of space. In the past, designers produced one or two design results by analyzing performance based on shape and materials. However, artificial intelligence can produce thousands of results in a short time considering various factors [5]. This design revolution in the manufacturing process will make the overall manufacturing process change.

5. Conclusion

This study reviewed changes in the manufacturing design process in the age of artificial intelligence. As these digital design technologies develop, the changes in design are as follows. After the designer has made a conceptual design, the designer enters the conceptual design in the Generative Design System. Then, based on the conceptual design, the computer provides the basic design. Then, if the designer chooses the appropriate design based on the basic design, artificial intelligence presents the optimal material proposal considering its function and economics. And the designer reviews this alternative. After the designer's review, the computer comes up with a final proposal. Much of this design is done by artificial intelligence. This is a design process that will come true in the future. The above-mentioned arguments of this thesis can be used as a theoretical basis for the manufacturing sector using artificial intelligence in preparation for the Fourth Industrial Revolution.

References

[1] Park Su Bin 2017 *The art of artificial intelligence age : study on the integrated curriculum of art with Google A.I. Experiments.* Gongju National University. Master's Degree Dissertation.

[2] Lee Jin Wook 2011 *A Study on the Characteristics of Furniture Design Using Generative Design - Focus on the Furniture Design using Fractal Geometry and Voronoi Diagram -.* Journal of Korean Institute of Interior Design. **84** Vol 20 pp 89-97

[3] Ahn Seong Mo etc. 2009 *A Study on Generative Design System through Ecological Concept Application.* Journal of Digital Design. **4** Vol 9 pp 279-288

[4] Jeon Young Jun 2018 *A Study on the implication of AI development for design work force : focused on social contribution and core competence of future designers.* Hongik University. Master's Degree Dissertation.

[5] Jeong Won Joon etc. 2018 *A Study on the Role of Designer in the 4th Industrial Revolution -Focusing on Design Process and A.I based Design Software-.* Journal of Digital Convergence. **8** Vol 16 pp 279-285

[6] Phillip Isola etc. 2018 *Image-to-Image Translation with Conditional Adversarial Networks.* Berkeley AI Research (BAIR) Laboratory, UC Berkeley. **26** arXiv:1611.07004v3 [cs.CV]

[7] Christopher hesse. 2017 *https://affinelayer.com/pixsrv/.* Accessed: 2017-04-21. 9

[8] Lee Jae Young 2018 *A Study on Construction of A.I.- Based Architecture Design Studio.* Hoseo University. Master's Degree Dissertation.

[9] Koo Bon Mee 2009 *A Study on the Characteristics of Generative Design on digital environment -Focus on Voronoi Diagram applied the Agent-based System-.* Journal of Basic Design & Art. **5** Vol 10 pp 15-23

Finite Element Study on Chain-Die Forming of an AHSS Channel with Variable Depth

F F M BENE[1,a], Z Y Liang[1,b], H Yang[1], D Y Li[1,c,*] and S C Ding[2,d]

[1]State Key Laboratory of Mechanical System and Vibration, Shanghai Jiao Tong University, Shanghai 200240, China
[2]School of Mechanical and Mining Engineering, University of Queensland, St Lucia,QLD 4072, Australia

Corresponding author: [a]franckbene@sjtu.edu.cn; [b]liangzhenye@sjtu.edu.cn; [c,*]dyli@sjtu.edu.cn; [d]s.ding@uq.edu.au

Abstract. A novel manufacturing technology called chain-die forming was in recent years proposed by Dr. Ding as an alternative to conventional roll forming. Mainly applied in the automotive industry, its prolonged deformation range enables the forming of advanced high strength steels (AHSS) with considerably reduced residual strain, which enables an adequate response to the increasing market requirements of this industry. In this paper, the finite element simulation of chain-die forming an AHSS channel with variable depth is performed. For more accurate predictions, the Bauschinger effect is well considered through the Chaboche hardening model implemented. The residual plastic strain and thickness distributions of the U shaped variable-depth profile are numerically evaluated. The forming limit at a critical zone is analyzed and the predicted magnitude of springback after forming is also calculated. The results portray the forming feasibility and strength of the chain-die forming technology as a valid alternative solution to stamping and roll forming, which are commonly used methods to manufacture such products. In one pass, a complete profile is achieved as opposed to roll forming, with no product defect such as wrinkling or cracking with limited springback development.

Keywords: Chain-die Forming, Advanced high strength steel, Variable-depth profile, residual plastic strain, springback

1. Introduction

Variable-depth products are commonly used in the automotive industry, some of which include frame rails, bumpers and door beams [1]. In order to meet the rising light weight, quality and safety demands of the automotive market, advanced high strength steels(AHSS) have become very popular in recent years in manufacturing companies. Constant profile products may be easily formed through roll forming or stamping, but as regards more complex shapes with variable width or depth, additional technique needs to be included. For example stamping AHSS products requires a servo-driven press, which with the help of a pressure sensor controls the slide speed all through the stroke in order to ensure the shape of the formed product[2]. It is a complex control process which sometimes produces defects such as wrinkling and cracking. AHSS high strength will also require a high press load which most often results in high springback due to reverse loading on the product. Conventional roll forming also has the limitation of only forming products with a regular cross section. In recent years flexible roll forming has been a feasible response to the limitations of traditional roll forming by introducing transversely moveable forming rolls [3]. The rolls can move along a desired path of bending to form products of variable cross-sections[4]. A numerical control system is used to command the rotational and translational displacements of the rolls. This is as well a complex forming process which usually has web-warping, end flaring and wrinkling as common defects[5].

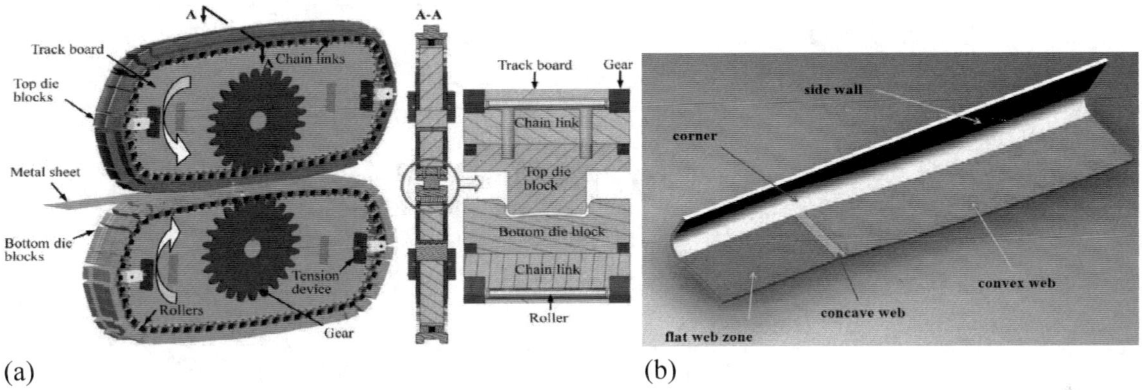

Figure 1. (a) Schematic diagram of the Chain-die former [6]; (b) quarter section feature definition.

In the above mentioned forming methods just like in most metal forming processes, product defect is as a result of redundant plastic strain development and residual stress producing springback. Chain-die forming(CDF) was first proposed by Ding et al [5] in 2008 as an alternative to roll forming. This forming technology boasts the advantages of low residual stress on the end product (due to an increased virtual rotation radius), insignificant redundant plastic strain, simplified control unit and cheaper tooling cost. In comparison to roll forming, the chain-die former increases the forming length to more than 10 meters [7] with a ratio of 1:100 [8, 9]. CDF is a combination of bending and stamping. Discrete die blocks are mounted on two track boards each having a large radius. The die blocks are held together by chain links and mounted as to rotate round the track board as shown in figure 1(a). The blank is fed incrementally in between the die blocks at a quasi-static rate in one pass to form the product.

Various studies have been carried out on forming variable width products by flexible roll forming and chain-die forming but forming variable depth products by CDF remains unexplored. This study seeks to investigate the feasibility of chain-die forming, to a variable depth product. For that, it is necessary to investigate the residual plastic strain after forming, the thickness distribution of the product as well as the springback. Two tooling corner radii are considered for this study.

2. Material and Product Dimensions

AHSS have proved to exhibit a high level of bauschinger effect when used in forming processes which involve loading and reloading[10]. The Chaboche hardening model has been proved to better describe the AHSS material behaviors during numerical simulations, especially springback[6]. Following previous work [11], the Chaboche hardening model is used to predict the residual plastic strain and springback. The Mises yield criterion is employed here as:

$$f = \| \sigma - x \| - \varphi = \left(\frac{3}{2} (\sigma' - x') : (\sigma' - x') \right)^{\frac{1}{2}} - \varphi = 0 \quad (1)$$

where $\varphi = \sigma_0 + \sigma_{\mathrm{ISO}}$, φ is the yield surface size in stress space, σ0 is the initial yield stress, σ_{ISO} is the isotropic hardening stress, σ' and x' are deviatoric parts of stress (σ) and back stress (x) respectively. The back stress and isotropic hardening parameters implemented as benchmark were obtained from Y. Li et al [10]. The Bauschinger ratio for the material is 0.811.

Table 1. Mechanical properties of the material (tested by a Zwick/Roell Z020 machine at SJTU)

Thickness (t)	1.2mm
Young's modulus (E_0)	204 GPa
0.2% Offset yielding stress (σ_s)	884 MPa
Ultimate tensile strength(σ_{UTS})	1403 MPa
Uniform elongation(e_u)	16.2%
Total elongation(e_t)	18.2%

The material used in this analysis is QP1180 provided by Baosteel. The mechanical properties of the material is as shown in table 1. Local features are defined on a quarter section of the product as shown in figure 1(b). The stress/strain curve is presented in figure 2.

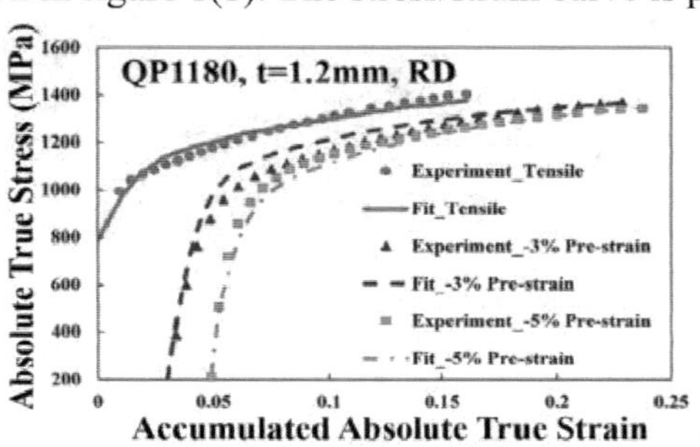

(a)

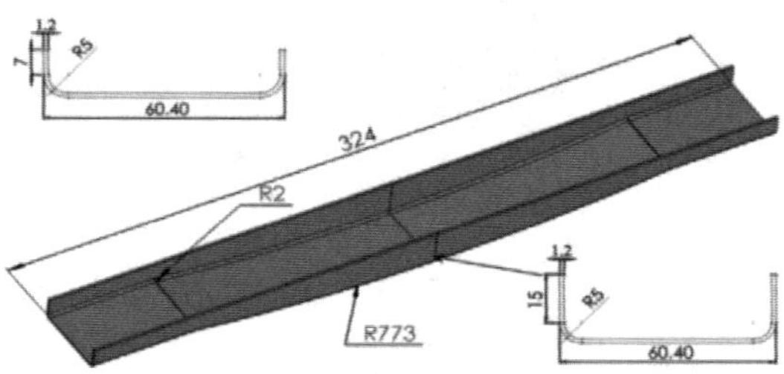

(b)

Figure 2. (a)Absolute true stress vs accumulated absolute true strain curve of QP1180[10]; (b) Part section and dimensions in mm.

3. Finite Element Modelling

ABAQUS/Standard is employed to carry out the chain-die forming simulations. Due to the symmetry, only a half section is considered for the simulation to save computing time. The dies are computed as rigid body parts and the blank as a deformable part. Both the top and bottom dies are split into 9 fragments in conformity with the chain-die forming machine. The blank is attributed a section of type solid-homogeneous. A large virtual radius of 35m is assigned to the top and bottom die pairs using reference points [10]. A surface to surface contact type is employed between the die and the blank. A penalty friction formulation is used with a friction coefficient of 0.15. A fit exponential curve is used as a pressure-overclosure to compute the normal behaviour during contact. The blank is given a very fine mesh for convergence purposes. The critical zones as previously defined in figure 1(b) are attributed an even more refined mesh to improve accuracy as shown in figure 3. The blank is meshed with 77056 linear hexahedral elements of type C3D8R. The dies fragments having a coarser mesh totally, are made up of 4446 linear quadrilateral elements of type R3D4. A symmetry boundary condition is applied on the blank as shown in figure 3.

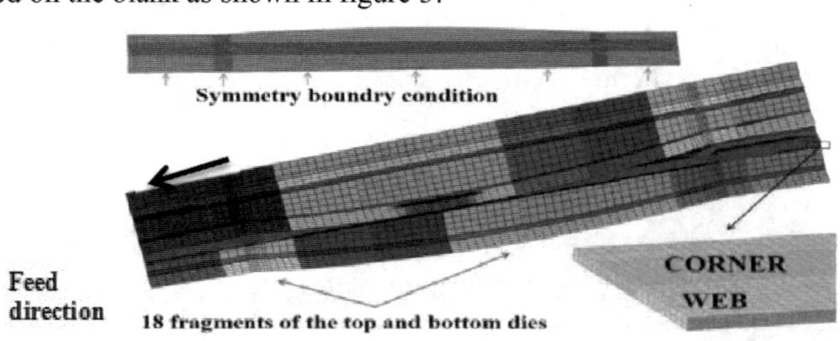

Figure 3. Assembly diagram of the meshed chain-die former with workpiece.

At the start of the simulation process, in order to avoid a blank distortion at the first contact between the blank and the die blocks, the blank is slightly stamped at a minimal level between the first block-pair of the chain-die former. This enables a tight grip for the incremental forming process to continue smoothly. During the forming process, the top die punches and bends the workpiece gradually upon the bottom die till the middle of the chain-die former. After the middle is crossed, the unloading process begins until both dies separate as the chain-die former rotates.

4. Results

The effective plastic strain, the forming limit at a critical zone, the thickness distribution along longitudinal paths and springback on a quarter section of the product are discussed in this section. In figure 4, Lines 1, 2 and 3 are used to represent section cuts performed to study the effective plastic strain at three regions on a quarter section of the product while paths 1, 2, 3 and 4 represent longitudinal nodes whose coordinates are extracted to evaluate the thickness distribution on the web, corner and sidewall of the same quarter section. Overall, the results portray good forming feasibility, the material properties are preserved as evidenced by negligeable redundant plastic strain and limited springback. Further details are described below.

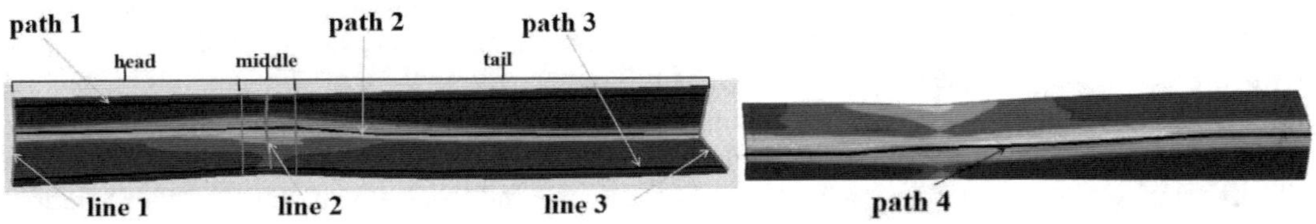

Figure 4. Transverse lines and longitudinal paths used to study the plastic strain development.

4.1. Effective plastic strain

The effective plastic strain across the thickness at the head, middle and tail zones are observed through section cuts along lines 1,2 and 3. Figure 5 primarily reveals that the strain concentrates are all found in the corner region of the blank. The highest strain concentrations are on the inner surface and outer surface of the blank corner. The maximum effective plastic strain developed at the outer corner surface of the head zone is 0.1444 whereas a maximum value of about 0.0642 is observed on the inner surface. Overall, the middle zone exhibits the maximum strain of 0.1925 on its inner region and 0.1765 on the outer region. This is normal as bending is simultaneous in the longitudinal and transverse directions in the concave region. Along the tail zone a maximum effective plastic strain of 0.1604 is also developed on its outer surface while its inner surface reveals a maximum value of about 0.0802. All three zones have a neutral middle corner layer which portrays no plastic strain. There is a negligeable amount of plastic strain in the web and sidewall of the middle zone, and almost no redundant plastic strain in that of the head and tail zones except for the deformation region (the corner). This therefore shows that no product defect such as wrinkling or cracking should be expected.

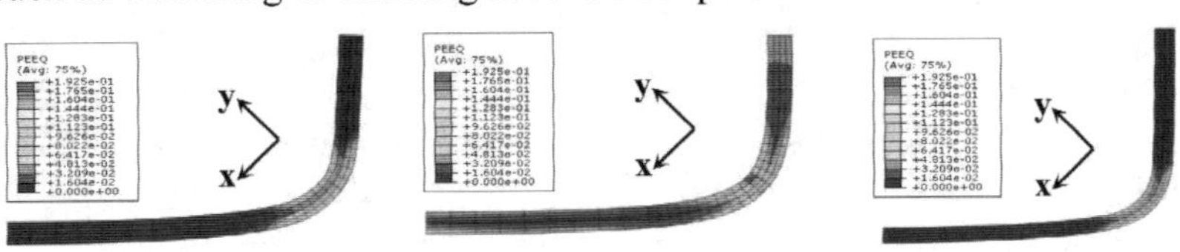

Figure 5. Effective plastic strain display in cross-section cut along lines 1,2 and 3 respectively.

4.2. Forming limit check at one critical region

One critical consideration for variable depth profiles which makes forming difficult for other forming methods is that the concave zone of the corner experiences both transverse and longitudinal loading simultaneously. A special attention is therefore given to that region. The major and minor strains of a group of elements in the concave corner as shown in figure 6(a) are extracted. By curve fitting, their

IOP Conf. Series: Materials Science and Engineering **727** (2020) 012012 doi:10.1088/1757-899X/727/1/012012

relationship is compared in a forming limit diagram to that of the work-piece. Two bending corner radii are considered, as shown in figure 6(b). The result shows that both bend corners radii of 1.2mm and 5mm can be formed by chain die forming without cracking.

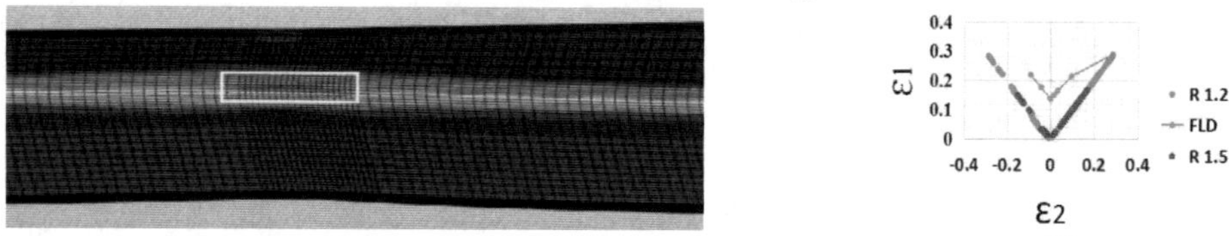

(a) (b)

Figure 6. (a) Selected region for element strain data extraction; (b) Forming limit diagram.

4.3. Thickness distribution along longitudinal paths

After the springback simulation, the displacement coordinates of nodes along longitudinal paths 1,2,3,4 are examined to investigate the thickness distribution of the blank after forming (see figure 7). For the head region, the results show a small thickness variation with the maximum difference being only about 0.17mm along path 3 at the middle surface of the web. The middle zone also shows even less variation with 0.13mm in the corner zone. This is normal since there is a high strain concentration found in the concave zone (see figure 6 (a)) resulting in the thinning of that area. The tail region towards its end along path 3 shows the greatest variation of 0.23mm, though still within an acceptable range. This is reasonable as it represents the middle convex part of the web, which experiences the highest load during forming. Greater springback is also expected from this area as shown in figure 8(a).

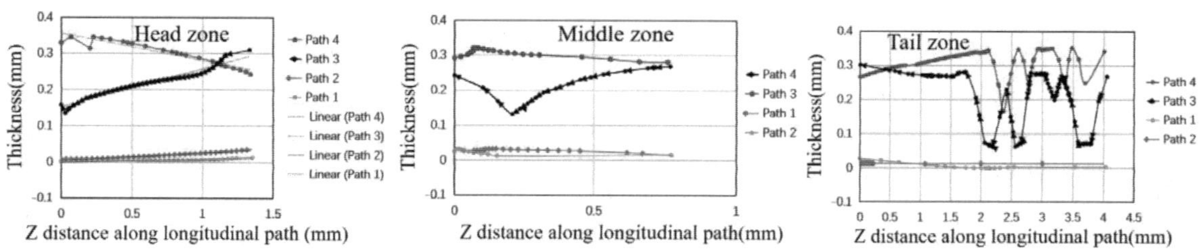

Figure 7. Plastic strain development along longitudinal paths in the head, middle and tail zones.

4.4. Springback development

One other product defect which needs to be studied is the springback which occurs during the unloading process mostly due to residual stresses in the product. The coordinates of longitudinal nodes along the middle section of the web bow and that of transverse nodes along line 2 are extracted. For the convex bow as seen in figure 8(a), a springback of about 2mm is observed in the convex web of the 1.2mm corner radius part as opposed to about 2.5mm on the 5mm corner radius part. This reveals a direct relationship between the corner radius and the web bow. The larger the corner radius the greater the web springback. A similar relationship is observed in the concave corner(see figure 8(b). There is a larger springback of only about 0.7 mm which occurs in the R5 corner in contrast to about 0.5mm in the R1.2 corner. Overall this low springback level could easily be compensated for by re-designing the die blocks in the future.

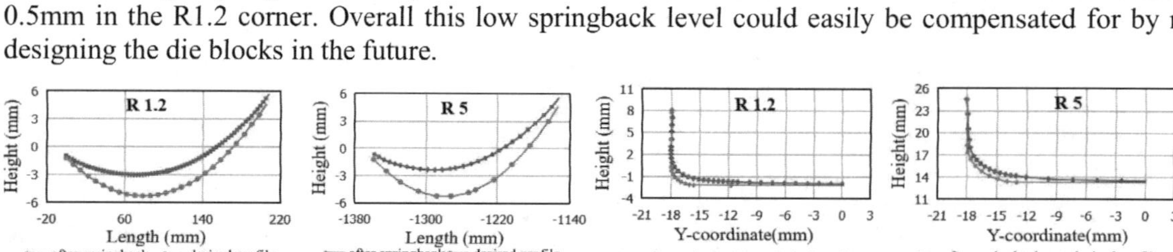

(a) (b)

Figure 8.(a) Springback comparism of entire convex zone; (b) Springback comparism at concave zone.

5. Conclusion and Remarks

A finite element study on chain-die forming a variable-depth product with emphasis on possible product defects which can be caused by residual plastic strain after forming and springback development is covered in this paper. The forming limit check at a critical area on the product as well as the thickness distribution on the formed product after springback is also discussed. The following conclusions can be drawn from this analysis:

- A variable depth product can be manufactured by chain-die forming in a single pass.
- The effective plastic strain and residual strains as well as the forming limit check, show that no obvious product defects, such as wrinkling, twisting and cracking, are to be expected on the product.
- There is a low amount of springback is in the corner region but a more noticeable amount in the web of the chain-die formed variable depth product.

Due to the accuracy of the springback predictions through the bauschinger effect included in the material's strain hardening behaviour, further springback compensation studies will also produce a more accurate product. Further studies are currently being carried out and an experimental validation will also be conducted to verify the residual plastic strain predictions and product quality.

Acknowledgments

The authors acknowledge the financial support by NSFC [Grant number U1860110]. The authors express their thankfulness to Baosteel who provided the material used in this work. The valuable inputs and comments from the colleagues in SJTU is also recognized.

References

[1] Joo B D, Han S W, Shin S G R and Moon Y H 2015 Flexible roll forming process design for variable cross-section profile *International Journal of Automotive Technology* **16** 83-88

[2] Abeyrathna B, Rolfe B, Pan L, Ge R and Weiss M 2016 Flexible roll forming of an automotive component with variable depth *Advances in Materials and Processing Technologies* **2** 527-538

[3] Kasaei M M, Naeini H M, Abbaszadeh B, Silva M B and Martins P A F 2015 Flexible roll forming, *Materials Forming and Maching* 51-71

[4] Larranaga J 2011 *Geometrical accuracy improvement in flexible roll forming process by means of local heating*, PhD Thesis (Department of Mechanical and Industrial Production, Mondragon University)

[5] Ding S, Daniel W, Yuan J and Zhang Y 2008 Making roll forming flexible-introduction to Chain forming *Yokohama Tube and Pipe 2011 Joint Symposium-Innovative Tube and Pipe Manufacturing and Forming: proceedings* 335-340

[6] Li Y, Sun Y, Xiao H, Bong H J, Shi L, Li S, Li D Y, Ding S C and Wagonere R H 2018 A numerical study on chain-die forming of the AHSS U-channel and contrast with roll forming *International Journal of Mechanical Sciences* **135** 279-293

[7] Li Y G, Huang H L, Li D Y, Sun Y and Ding S C 2017 Finite Element Simulation of Chain-die Forming U Profiles with Variable Cross-Section *Materials Science Forum* **898** 1177-1182

[8] Lindgren M 2007 Cold roll forming of a U-channel made of high strength steel *Journal of Materials Processing Technology* **186** 77-81

[9] Bidabadi B S, Naeini H M, Tehrani M S and Barghikar H 2016 Experimental and numerical study of bowing defects in cold roll-formed U-channel sections *Journal of Constructional Steel Research* **118** 243-253

[10] Zang S, Sun L and Niu C 2013 Measurements of Bauschinger effect and transient behavior of a quenched and partitioned advanced high strength steel *Materials Science & Engineering A* **586** 31-37

[11] Chaboche J and Rousselier G 1983 On the plastic and viscoplastic constitutive equations - part I: rules developed with internal variable concept. *Journal of Pressure Vessel Technology* **105** 153-158.

IOP Conf. Series: Materials Science and Engineering **727** (2020) 012013 doi:10.1088/1757-899X/727/1/012013

Study on the Correlation between the Texture Depth of Volume Replacement Method and the Texture Depth of Spectral Reproduction Method

Zhang Jinning[1], Zhou Yishu[1,*], Wang Pengfei[1] and He Huayang[1]

[1]Research Institute of Highway Ministry of Transport No. 8 Xitucheng Road, Haidian District, Beijing

[*]Corresponding Author E-mail: ys.zhou@rioh.cn

Abstract. Through theoretical research and actual road condition analysis, this paper establishes the simulation calculation model and the spectrum analysis model. Through the spectrum inversion, the stable road surface state is obtained, and the conversion relationship between the reconstruction depth of the spectral reproduction method and the structural depth of the volume displacement method is obtained. Through the conversion relationship of a typical single spectrum, the relationship between SMTD and TD is obtained under each type of standard waveform. Actually, in the process of sand paving, due to the presence of the exposed part, the SMTD and TD in the actual road surface are not positively correlated, thereby correcting the relationship between SMTD and TD. The acquired road surface data is modeled by MATLAB, and the analog waveform is made according to the actual waveform, and the reduced order decomposition is performed to determine the range of a. The intercept b value is determined by comparing the experimental spectra before and after sanding. This paper solves the problem that the detection mechanism needs to perform statistical correlation test in accordance with the requirements of JTG-E60 regulations to meet the requirements of pavement assessment.

1. Introduction

The pavement texture depth is a technical indicator of the surface performance of the pavement and is one of the important technical parameters for measuring the anti-sliding performance of the pavement. There are two main types of measurement methods for structural depth. One is the contact measurement method based on the volume conversion principle, that is the texture depth of volume replacement method. The other type is the non-contact measurement method based on the cross-section measurement, that is the reconstruction method of the spectral reproducibility method, which is represented by the laser texture depth meter test method. China's transportation industry standard JTG F80/1-2004 highway engineering quality inspection standard evaluation method is the texture depth of volume replacement method. The texture depth of spectral reproduction method, that is, the texture depth of laser method, is not included in the standard. With the development of electronic technology, the application of vehicle-mounted laser depth gauges has become increasingly widespread. JTG E60-2008 highway subgrade pavement field test procedures require that the correlation coefficient between the texture depth of laser method and the texture depth of volume replacement method is not less than 0.97 [1]. Therefore, the laser texture depth test must be tested by the volume replacement method. Studying the relationship between the texture depth of spectral reproduction method and the texture depth of volume replacement method can greatly reduce the workload of construction and inspection units.

This project has mastered the technical points of the correlation between the texture depth of volume replacement method and the texture depth of spectral reproduction method. It basically overcomes the technical difficulties of the original output parameters being non-uniform and non-correlated, and realizes various calculation methods of structural depth. Unification provides a reference for the detection and acceptance of structural depth, and provides a technical basis for government departments to formulate policies.

The establishment of the texture depth of spectral reproduction method and the texture depth of volume replacement method indicates that China can also quickly evaluate the pavement by the spectral reproduction method. Because the texture depth of laser method in China has a complete traceability system, China's structural depth measurement has reached the international leading level[2].

2. Research content

In the field test procedure of JTG E60-2008 highway subgrade pavement in China's transportation industry standard, there are three kinds of test methods for pavement depth: manual sanding method, electric sanding method and vehicle-mounted laser depth gauge method. The first two are contact measurement methods, and the latter one is non-contact measurement method. There are two types of non-contact pavement texture depth calculation models: pavement texture depth measurement technology based on the spectral theory and the depth measurement technology based on the section elevation pavement structure.

2.1. Computational model of pavement texture depth measurement technology based on spectral theory

The internationally recognized definition of pavement depth is the deviation of the actual road surface profile from the ideal (reference) plane. In three-dimensional space, the specific point of the road surface is within a certain area (usually the same size as the tire grounding area). The distance (TD) between the ideal planes passing through the three highest points.

Although the ideal plane in the definition is difficult to form in actual measurement, it is more measurable than the definition of pavement depth in China. There is a certain degree of difference between the original data filtering and texture depth calculation models by various manufacturers, especially the texture depth calculation model. At present, there are mainly two types of texture depth calculation models, one is from the International Organization for Standardization (ISO) ISO13473, MPD calculation model; the other is from the British Institute of Transport and Road Research, 1974 Technical Document No. 639, SMTD calculation model.

China's original pavement laser structure depth gauge was introduced from abroad, and its corresponding calculation model is also from abroad. With the different origins of the instruments, the calculation models used are quite different. Some adopt British BS standards, and some adopt ASTM and ISO standards, some even refer to the same standard system, due to the specific technical methods selected and the special parameters used, the calculation results are also very different. For example, ISO13473 proposes four calculation methods of MTD, PD, MPD and ETD, while ASTM E1845-96 also describes MPD and ETD calculation methods, and UK SCANNER measurement system provides SMTD, MPD and RMST. Calculation method. Due to the difference of the calculation model, it is very likely that the same section data will appear and different texture depth values will be obtained. This results in low reliability and poor comparability of the pavement texture depth detection data in China, which has seriously affected the evaluation of the anti-sliding performance of pavement in China[3].

China GB/T26764-2011 "Multi-function road condition rapid detection equipment", the relevant provisions on the pavement structure depth equivalent to the TRRL method, which does not give its value traceability method, but far from the uniform value, standardizing the production purpose of instruments and equipment. Australian equipment manufacturer ARRB uses a gear set to perform the laser calibration of the depth gauge factory calibration test. The vertical installation rotation is unstable, the gear replacement is difficult during testing, and the traceability trace chain is not clear. Due to the difference of calculation models and the lack of uniform value traceability method, the evaluation of road surface anti-sliding performance is not at the same level, which affects the scientific rationality of road quality evaluation. ISO, ASTM and related well-known scientific research institutions all define the pavement structure from the perspective of road surface spectrum. That is, the normal road surface longitudinal section is considered to be a set of continuously distributed spectra after the Fourier component decomposition. This spectrum can be It is quantified by different wavelengths, and the pavement structure is a set of spectra with a specific wavelength range[4].

2.2. Computational model of pavement texture depth measurement technology based on section elevation

The elevation measurement technology of pavement structure based on elevation is based on the theory of pavement spectrum. Firstly, the laser ranging technology is used to measure the elevation of the pavement profile quickly and accurately. Then, a certain calculation model is used to evaluate and describe the wavelength range of 0.5-50mm. The geometry of the road surface inside, thereby solving the measurement of the pavement texture depth. There are many different ways in the world to evaluate and describe the calculation model of pavement texture depth.

Profile Depth, PD: In a two-dimensional space, the distance between a point in the longitudinal section and the horizontal line of the highest point in the section in the interval of the tire grounding length.

Mean Profile Depth, MPD: The average of the depth of the section over a specific length interval.

Estimated Texture Depth, ETD: Using MPD as an independent variable, the conversion formula (such as ETD=0.2+0.8*MPD) is used to evaluate the depth of pavement structure.

Sensor Measured Texture Depth, SMTD: For the height series of the test points in a certain length paragraph, the calculated residual after the quadratic parabolic regression is performed. As shown in Figure 1.

$$\text{SMTD}=\sqrt{\frac{\sum_1^N(y_i-y_e)^2}{N}} \quad (1)$$

Among them: SMTD—the texture structure depth index, parabola regression is performed on the height data of each detection point, and the residual can be obtained. The unit is mm.

y_e—a parabola obtained by quadratic least squares regression, which is $a+bx+cx^2$.

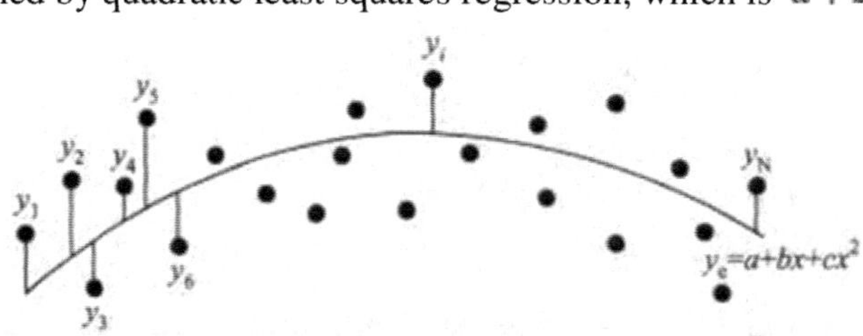

Figure 1. Sensor calculation model for measuring texture depth.

For the construction depth meter, regardless of the texture depth of volume replacement method or the texture depth of spectral reproduction method, the domestic and foreign technologies are relatively mature, but in terms of output parameters, there is no correlation formula of each output parameter. The study of the correlation between the texture depth of volume replacement method and the texture depth of spectral reproduction method will further develop the laser structure depth detecting equipment and solve the problem that has not been solved in the industry for a long time.

3. Main research method

3.1. Theoretical study on texture depth conversion method

The working principle, input and output of the depth device were constructed by the spectral reproducing method to construct the depth and volume displacement method. The two parameters of the texture depth were analyzed according to the spectral theory. On this basis, the typical waveform is tested and the theoretical conversion formula of two parameters is obtained.

3.2. Test of texture depth conversion method

Firstly, the theoretical model is established, and the calculation software of different algorithms is output through MATLAB, excel and various construction depths to perform calculation and verification, and the relationship or correlation of output values of various construction depth devices is determined [5]. After completion, various types of asphalt surface materials and various road surfaces across the country are collected, modeled and calculated according to the sampled data, and the data is reproduced. Finally, the spectral data is reproduced by the scanned data to obtain a conversion formula. The main implementation plan and technical route of the project are shown in the figure.

4. Theoretical research

4.1. Brief description of spectral analysis methods

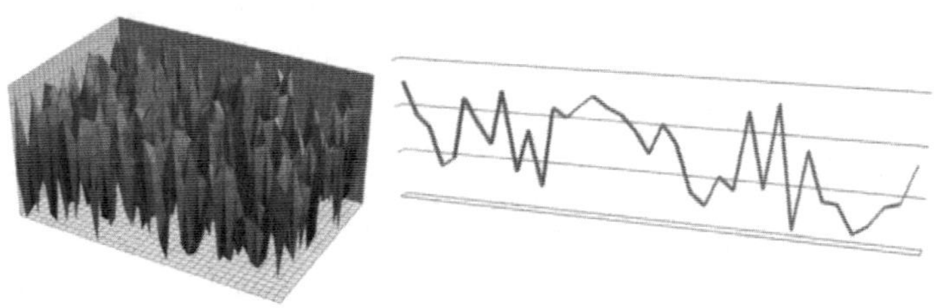

Figure 2. Collected part of the actual road surface data and Two-dimensional spectrum of acquired data.

ISO, ASTM, and well-known foreign scientific research institutions (such as TRRL) define the road surface structure from the perspective of road surface spectrum. That is, the normal road surface longitudinal section is decomposed by Fourier components and will contain a set of continuous distribution. Spectral spectrum, which can be quantified by different wavelengths, and the pavement structure refers to the deviation of the actual road surface longitudinal section from the ideal plane. Permanent International Association of Road Congresses has defined pavement construction in different wavelength ranges, which is widely recognized by the international engineering community. Since the volumetric displacement method is a computational model that aggregates points into three-dimensional measurements, the depth of the spectral reconstruction method is two-dimensional. The original research mainly uses statistical correlation, and no mathematical methods are used for research. In this study, the three-dimensional magnitude is differentiated to obtain the depth value of the two-section sanding method, and the relationship between the two is found. As shown in Figure 2 and Figure 3.

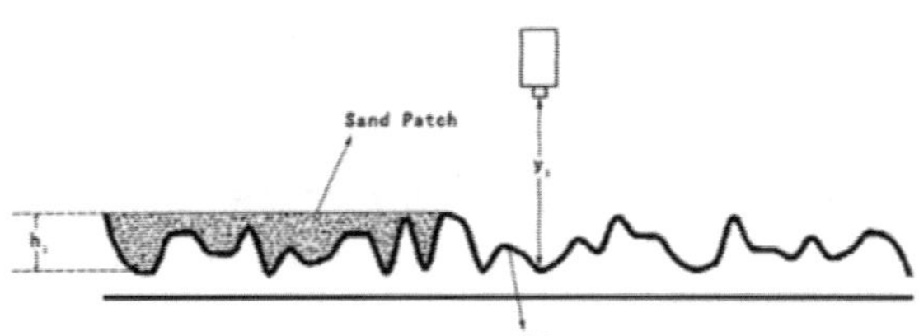

Figure 3. Section of the texture depth of volume replacement method and the texture depth of spectral reproduction method.

4.2. Conversion calculation of typical single spectrum

This paper simulates the value of a variety of basic waveforms and waveform changes. The total count value is 7 types, 116 groups, and nearly 10,000 waveform data[6]. One set of data is shown in Figure 4.

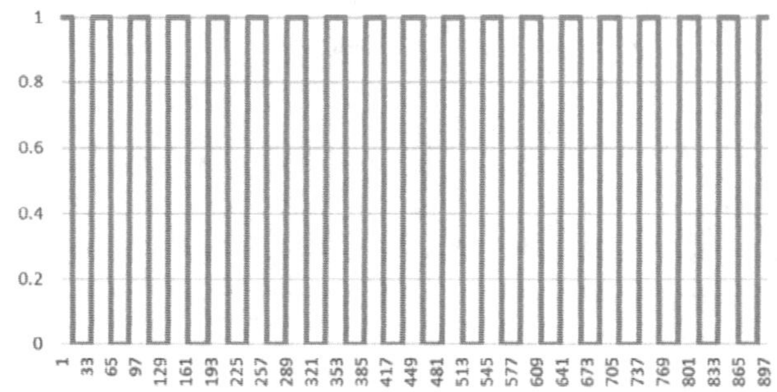

Figure 4. Square wave basic waveform data - Simulated cement concrete groove.

This paper is calculated by the standard waveform input software. Under the standard waveform, the relationship between each type of SMTD and TD is SMTD=TD*a(a>0). The SMTD and TD of each type of waveform are positively correlated, and there is only one correlation coefficient. The wavelength and amplitude of the waveform below 50 mm do not affect the value of the correlation coefficient. The correlation coefficients of different types of waveforms are different.

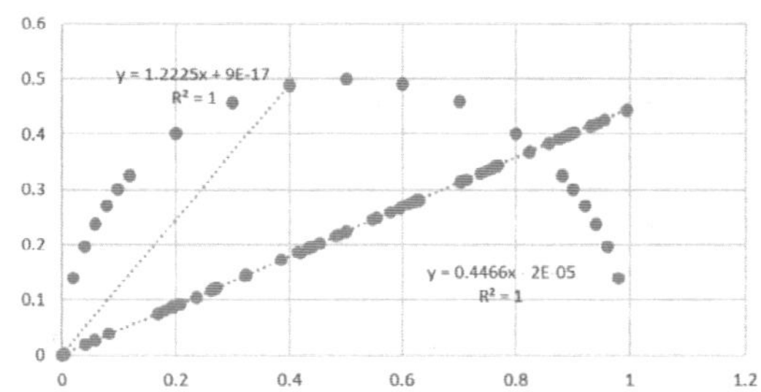

Figure 5. SMTD and TD relationship diagrams of different standard waveforms

Select typical data as shown in Figure 5. The red line represents the line type of one of the waveform data. The waveforms of different values and parameters are shown by the red dot in the figure. After the calculation, the intercept is extremely small and negligible. The blue dots represent different types of waveforms, and the corresponding a values are different. Each blue dot can be linearly the same as the gray one. This conclusion is consistent with the linear correlations between SMTD and TD made by various institutions in the world over the years. It also reveals that for many years, due to the uncertainty of a value, there has not been a correlation formula to convert the relationship between SMTD and TD [7].

5. Experimental research

According to the wave theory, any waveform is superposed with different waveforms, that is, the road waveform can be decomposed into waveforms of various waveforms [8]. The correlation coefficient a of the TD value and the SMTD value is also completed in a different waveform synthesis superposition. Although the value of a is different on different road surfaces, the waveform of the road surface is not extreme, and the data has certain regularity [9]. There is a range and interval. In order to determine the a-value interval, this paper has verified the theoretical and practical experiments.

5.1. Pavement waveform decompression calculation

In this paper, MATLAB is used to model the acquired pavement data, and the model is decomposed and reduced. At the same time, the simulated waveform is made according to the actual waveform, and the reduced-order decomposition is performed[10]. As shown in Figure 6.

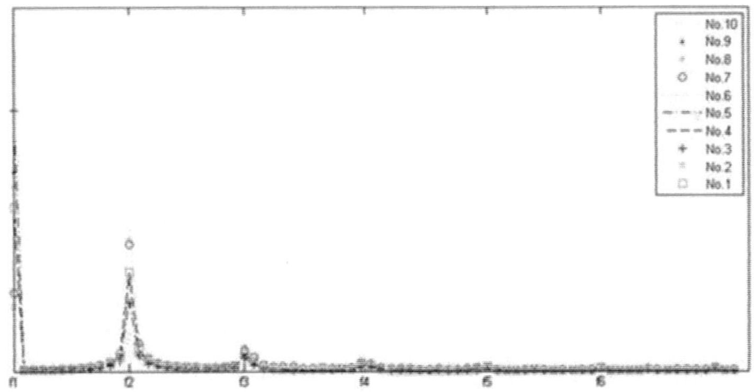

Figure 6. Spectrogram of different pavement models.

The relationship between SMTD and TD after degrading the pavement waveform is shown in Figure 7. The horizontal axis of the coordinate system is the frequency of the wave, and the vertical axis is the amplitude[11]. According to the calculation results of different decomposition data, the value of a is basically in the range of 0.4~0.6.

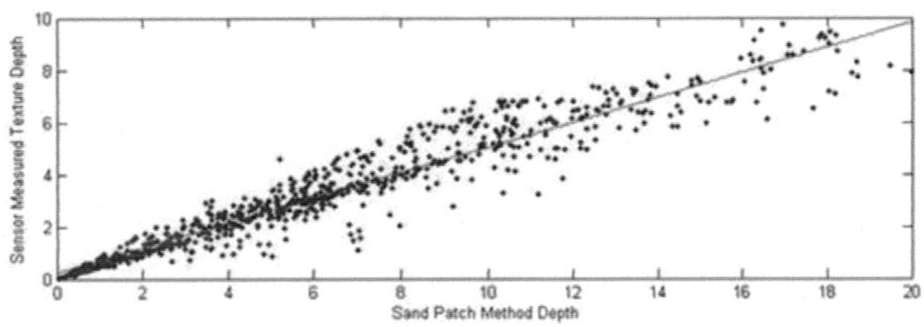

Figure 7. Relationship between SMTD and TD after degraded decomposition of pavement waveform.

5.2. The difference between the actual road surface waveform and the theoretical waveform

This paper collects asphalt surface data and actual road surface data through 3D scanner, collects and analyzes the data, and determines the range of a value[12]. The road surface and analog disk results scanned by the 3D scanner are shown in Figure 8.

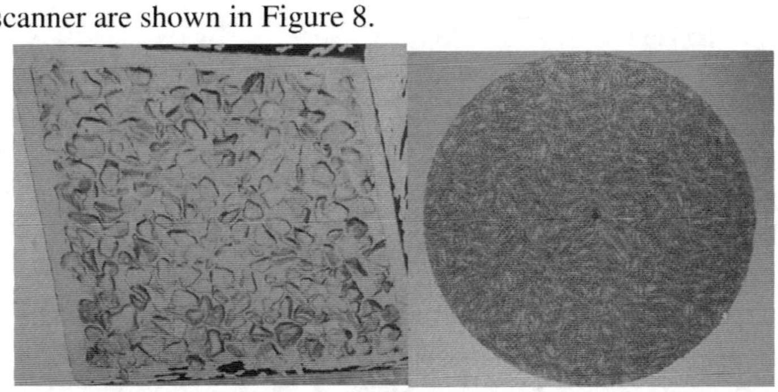

Figure 8. 3D scanner scanned image of pavement and analog disk.

Actually, the asphalt pavement cannot be completely covered by sand during the sanding process, and its partial height value is large, as shown in Figure 9. Due to the presence of exposed parts, SMTD and TD are not positively correlated in the actual road surface [13]. The SMTD still exists after the sanding section, so the conversion formula is not SMTD=a*TD in the actual process, but there is an intercept. That is, SMTD=a*TD+b.

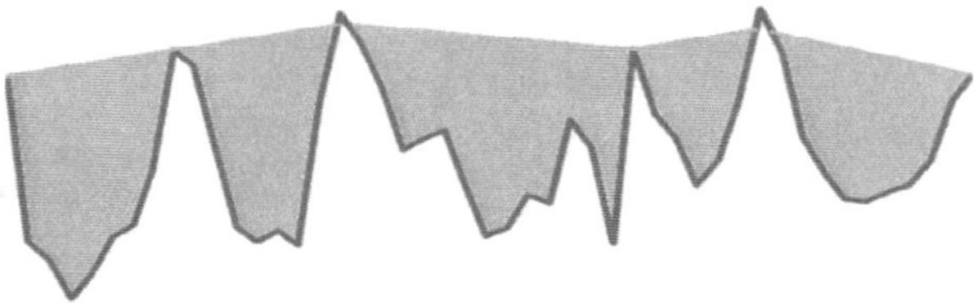

Figure 9. Schematic diagram of paving sand exposed.

5.3. Comparison of actual sanding and sanding theoretical values

In this paper, seven different technicians are selected to calculate the sanding of the standard disc. After each sanding, the standard disc is brushed clean and the next technician will sand again.

The experimental results show that the errors measured by different people are large during the sanding process[14]. After measuring 25 ml of sand with a balance, the density was weighed and found to have a density of 1.42, which is in line with the density range of standard sand. The theoretical sanding quality is obtained by converting the surface sand according to the density of the standard sand. Finally, the theoretical sanding results are compared with the actual sanding results, and the theoretical calculation values are basically consistent with the measured values [15].

5.4. Intercept Calculation

COOPER D R pointed out in Report No. 639: Measurement of Road Surface Texture by a Contactless Sensor that SMTD has a good linear correlation with TD, which is the main reason why the calculation model is widely adopted. The report obtained a correlation formula through a certain number of experiments: SMTD = 0.59 * TD + 0.12, and the correlation coefficient (R) was greater than 0.95[16].

Gan Baozhu, Zhang Yuguang, in the "Public Engineering" published "correlation between the sanding method and the laser structure depth gauge method to determine the depth of the pavement structure", proposed different correlation formulas and correlation coefficients. Under the condition of confidence a=0.01, the critical correlation coefficient of 96 groups of values is 0.2540, and the correlation coefficient is as high as 0.890. The formula can be used to convert the depth value of the spectral reconstruction method into the volume displacement method. The correlation formula is DTL=0.041+0.528DTS, where DTS is the volume displacement method texture depth, and DTL is the spectral reconstruction method texture depth, SMTD = 0.528*TD+0.041.

Table 1. SMTD value before and after sanding

SMTD value before sanding	SMTD value after sanding	Sanding part SMTD value
0.310	0.073	0.295
0.440	0.103	0.418
0.344	0.079	0.327
0.876	0.086	0.831
0.405	0.093	0.384
0.867	0.098	0.822
1.059	0.091	1.004
0.766	0.104	0.727
0.147	0.053	0.139
0.845	0.101	0.802

In this paper, the test blocks were measured before and after sanding, and the SMTD was measured as shown in Table I. Convert the contents of the table to a graph where the horizontal axis is the measured length and the vertical axis is the elevation value. The red line is the surface spectrum before sanding, the blue line is the blue spectrum after sanding, and the green line is the spectrum of the sanding part[17]. According to the measurement, the basic residual of SMTD after sanding is slightly less than 0.1mm, taking a certain insurance factor, taking b as 0.1mm[18].

According to the theoretical and actual measured values, the intercept b value is taken as 0.1 mm.

5.5. Calculation of Slope in Conversion Formula

In this paper, the test vehicle is used to test typical road sections, and ten typical sections are collected, totaling more than 450,000.

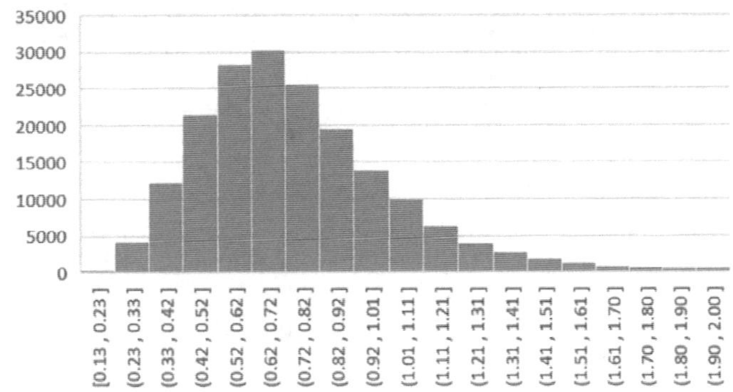

Figure 10. Distribution map of typical pavement surface value a.

As shown in Figure 10, In the distribution map of the typical asphalt surface value a, the horizontal axis represents the range of each interval of the value a, and the vertical axis represents the number of distributions of each interval.

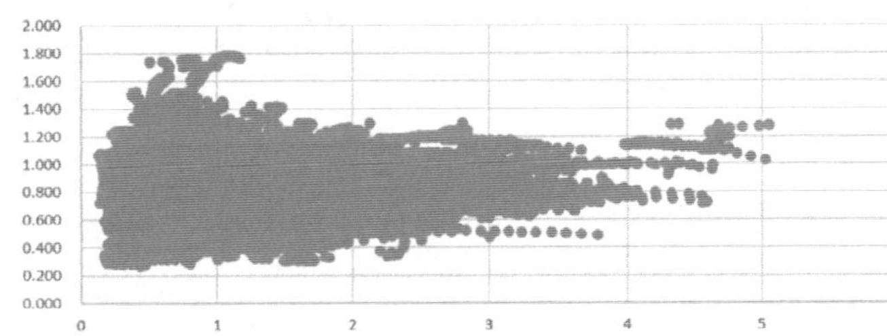

Figure 11. Distribution of SMTD and TD on typical asphalt surface.

Figure 11 shows the distribution of SMTD and TD on a typical asphalt surface. The horizontal axis represents TD and the vertical axis represents SMTD.

5.6. Conversion Formula

According to the above conclusion, the value a is 0.55 and the b value is 0.1, so the conversion formula is SMTD=0.55*TD+0.1.The confidence interval is 40%.

6. Conclusion

In this paper, the three-dimensional magnitude is differentiated by the spectral analysis method to obtain the depth value of the sanding method of the two-dimensional section, and the relationship between the three-dimensional and two-dimensional pavement structure depth measurement is determined. Through the calculation of the conversion relationship of a typical single spectrum, it is concluded that the relationship between SMTD and TD is SMTD=TD*a(a>0) under each type of standard waveform. Actually, in the process of sand paving, due to the presence of the exposed part, the SMTD and TD in the actual pavement are not positively correlated, and the SMTD still exists after the sanding section, so the conversion formula is SMTD=a*TD+b. The acquired pavement data is modeled by MATLAB, and the model is degraded. At the same time, the simulated waveform is made according to the actual waveform, and the reduced-order decomposition is performed. The range of a is determined to be in the range of 0.4~0.6, which is 0.55. By comparing the experimental spectra before and after sanding, it is determined that the intercept b has a value of 0.1 mm. In summary, the relationship between SMTD and TD is SMTD=0.55TD+0.1.

Acknowledgments
I would like to express my gratitude to all those who helped me during the writing of this paper. I gratefully acknowledge the help of my colleagues, who have offered me valuable suggestions in the academic studies. In the preparation of the paper, they have spent much time reading through each draft and provided me with inspiring advice. Without their patient instruction, insightful criticism and expert guidance, the completion of this paper would not have been possible. I should like to express my gratitude to my beloved parents who have always been helping me out of difficulties and supporting without a word of complaint.

References
[1] Dou GW 2014 *J. Traceability of Non-contact Pavement Structure Depth Value Journal of Chang'an University* **34** 6
[2] Masad E 1999 *J. Internal structure characterization of asphalt concrete using image analysis* **35** 7
[3] Dou GW 2015 *J. Research on Calculation Model of Pavement Structure Depth Based on Section Elevation Journal of Highway and Transportation Research and Development* **32** 1
[4] Wang DY 2004 *J. Evaluation and Measurement of Surface Depth of Asphalt Road Using Digital Image Technology Journal of South China University of Technology* **32** 2
[5] Zhu ZH 2014 *J. Application of grey feature weighted ELM in tectonic depth prediction Journal of Hefei University of Technology* **2** 2
[6] Chi FX 2008 *J. Evaluation Method of Surface Segregation of Asphalt Pavement Based on Laser Texture Instrument China Journal of Highway and Transport* **21** 26
[7] Shen W 2008 *J. Method and data processing method for measuring depth of structure by laser range finder* **22** 28
[8] Di Hanson 2004 *J. Evaluation of circular texture meter for measuring surface texture of pavements* **26** 282
[9] NZ L 2006 *J. Road surface texture measurement using digital image processing and information theory* **34** 22
[10] Gan BZ 2006 *J. Correlation between sanding method and laser structure depth meter method for measuring the depth of pavement structure* **4** 24
[11] Lu PF 2019 *J. Prediction of asphalt pavement wear condition based on pavement structure depth World of Transportation* **24** 37
[12] Teng JL 2019 *J. Research on prediction model of structural coal distribution based on deep confidence network China University of Mining and Technology* **27** 31
[13] Yuan SW 2019 *J. Yuan Shiwei. Depth Detection Technology for Expressway Asphalt Pavement Journal of Transportation* **45** 36
[14] Wang HY 2019 *J. Prediction of Wear Status of Asphalt Pavement Based on Pavement Structure Depth* **35** 30
[15] Yuan S 2018 *J. Evaluation of Loose and Polished Degree of Asphalt Surface Layer by Using Structural Depth* **05** 98
[16] Sun H *J. Correlation between structural depth and pendulum of SMA asphalt surface layer* **05** 96
[17] Ji ZY *J. Research on Depth Detection Technology of Asphalt Pavement Structure* **89** 80
[18] Liu J *J. Evaluation of Uncertainty of Pavement Structure Depth by Manual Sanding Method* **15** 14

IOP Conf. Series: Materials Science and Engineering **727** (2020) 012014 doi:10.1088/1757-899X/727/1/012014

Design and Research of Convenient Cantilever-Unloading Bracket for Antenna

Chunsheng Yang[1,*], Yiliu Xu[1], Yi Lu[1], Yu Wang[1], Feng Xue[1] and Zhenyue Ren[1]

[1]Beijing Institute of Space Environment Engineering, Beijing, China

[*]yangcs426@163.com

Abstract. In spacecraft assembly integration and test, zero gravity unloading of antenna is necessary during antenna deployment/closure test. To meet the requirement of antenna unloading condition in EMC test stage, a convenient cantilever type universal antenna unloading bracket is designed. The system consists of the main body of the cantilever device, the first-stage arm, the second-stage arm and the pulley system. Through analysis and calculation, the system can meet the space requirement, friction resistance requirement, strength and deformation limitation in the unloading process. It can provide technical reference and basis for zero gravity deployment/closure test for various surface antennas of spacecraft in the assembly process.

1. Introduction

In the process of spacecraft assembly integration and test(AIT), the deployment and closure tests of antennas are needed. At present, the fixed unloading truss-frame is used in the deployment test of antenna. [1]When the spacecraft carries out Electro Magnetic Compatibility (EMC) test, it is requested to deploy antenna in EMC test room. However, the space of EMC laboratory is not enough to set up fixed unloading truss-frame. At the same time, metal truss brings great interference to the wireless test signal during EMC test. [2]In the process of using the traditional fixed unloading truss-frame, it is necessary to assemble the truss, release the residual assembly stress, then transfer the spacecraft beneath unloading truss, install hanging device on the guide rail of the unloading truss, and unload the antenna in zero gravity. When switching to the antenna on the other side of the spacecraft for deployment/closure test, it is necessary to change the position of the spacecraft. It means the unloading truss-frame can't be moved while spacecraft should be repeatedly adjusted its attitude. Meanwhile, it takes long time to build up and truss-frame whit more cost and larger occupied space. [3] Such assembly mode leads to complicated operation with long assembly testing time, low efficiency, and small effective safety distance between truss-frame and spacecraft, which brings inconvenience to the operation of personnel and the layout in test room. [4]

Therefore, it is necessary to develop a kind of universal unloading bracket which can be easily moved. It is not only suitable for the deployment/closure of antennas in the EMC test stage, but also can provide stable structural support for the deployment/closure test of various antennas in other test stages, which makes the unloading operation of antennas more convenient. It can move the universal unloading bracket of antennas instead of moving the position of spacecraft. So it cost less time of spacecraft AIT and reduce the operational risk. [5] [6]

2. Structure design of convenient cantilever-unloading bracket for antenna

2.1. Overall Program

To meet the demands of deployment/closure of most spacecraft antennas, a large-span convenient cantilever-unloading bracket for antenna is designed. As shown in figure 1, the universal unloading bracket of the antenna adopts the form of cantilever type. The cantilever part is installed at the top of the main structure, equipping with the hanging mechanism of the air-floated pulley group.

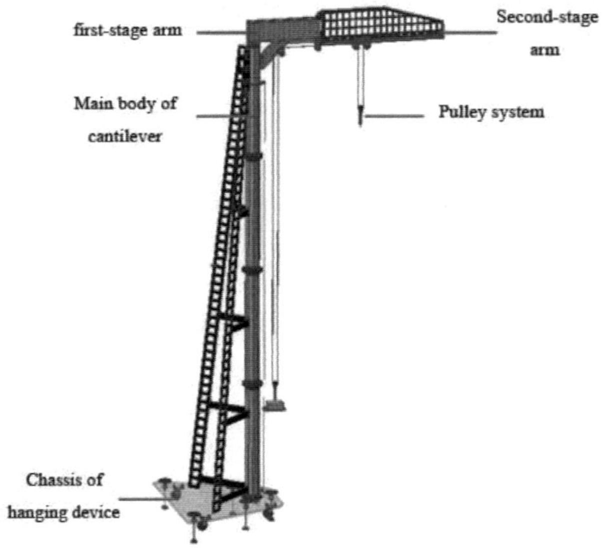

Figure 1. Overall drawing of convenient cantilever-unloading bracket for antenna

The main structure is composed of braces and chassis which can be assembled repeatedly according to the height requirement. The chassis is equipped with counterweight, mobile casters and adjustable supporting casters, which provides the stability of the system and has the functions of parallelism adjustment and overall movement. The cantilever part is mainly composed of a first-stage arm, a second-level deploying arm and a rotating axis; the first-stage arm is firmly connected with the main structure of the antenna universal unloading support; the second-level deploying arm can rotate freely around the first-stage arm; the bottom end of the second-stage deploying arm provides sliding guide. The fixed pulley assembly in the air-floated pulley group is fixed in the area of the second-stage arm, the first-stage arm and the counterweight box. The slider is installed on the guide rail of the second-stage arm and can slide freely along the guide rail. The movable pulley assembly is connected under the slide block through the Kevlar pull rope, and the end of the movable pulley assembly is a cylindrical suspender, which serves as the unloading interface for the antenna (or can be changed to other interface shapes according to its requirements). The size of unloading force is adjusted between pulley groups by Kevlar rope, counterweight box and air floatation mechanism; the counterweight box and Kevlar rope are fixed and guided by the side wall of the main strut to realize the follow-up of the counterweight box along the vertical direction; the whole set of pulley groups is air floatation structure, which can effectively reduce friction and realize zero-gravity unloading of suspended products. In the process of antenna deployment/closure rotation, the position of the hook can be automatically adjusted according to the height of the product. The movable pulley assembly slides along the guide rail and moves up and down through the pulley group. The second-stage deploying arm rotates around the first-stage deploying arm, realizing the coverage of the three-dimensional space.

The antenna universal unloading bracket has the advantages of convenient movement, flexible adjustment, less space resource occupation without limitation of site conditions. It can realize zero gravity unloading of products at any position. After the antenna is deployed and fixed, the antenna universal unloading bracket can be removed without affecting other test performances of spacecraft and other products. In addition, the system has sufficient stability, simple operation, and effectively avoids the potential safety hazards caused by operating in the tight space around the spacecraft. At the same time, it can meet the requirements of coverage of unloading area in three-dimensional space, and is suitable for most spacecraft antenna deployment/closure test requirements. It can completely replace other existing antenna unloading truss-frames.

2.2. Structural Constitution

The main body of the system is supported and fixed by the chassis of the hanging device which is equipped with supporting feet and universal rollers to adjust the level of the chassis and move the system on the ground, respectively; the supporting chassis is equipped with supporting columns, and the upper end of the column is fixed with a first-stage arm; the end of the first-stage arm is equipped with a second-stage arm through bearings at both upper and lower ends, and the second-stage arm can be rotated around the first-stage arm. The rotation range is larger than 90 degrees; the lower end of the rocker arm is equipped with linear guide rail; part of the pulley group is installed on the slide block of the guide rail to realize zero gravity unloading of the unloaded antenna, which can provide lifting force and a certain distance of lifting height. The hanging point position can be adjusted automatically with the change of the connecting point position.

1) Chassis of hanging device

The chassis of the hanging device provides the installation interface of the main body of the antenna unloading device. It also provides the installation position of power supply, control cabinet, etc. At the same time, the cantilever device is equipped with counterweight to ensure the stability of long-span load and anti-overturning of the unloading device. The lower end of the chassis is equipped with supporting feet and universal rollers, which can adjust the level of the chassis and move the system on the ground respectively. The shape of the chassis mainly refers to the avoidance relationship between the antenna support device and other mechanical equipment.

2) Main body of cantilever device

The main body of the cantilever device is comprised of 4 stainless steel tubes which can be assembled repeatedly. Each section of steel pipe is equipped with hanging points for assembling. Each section is connected by flange, and the chassis is fixed by flange and steel pipe reinforcement. According to the requirements of different working conditions, the total height of the antenna unloading device can be changed by reducing the number of steel tubes.

3) First-stage arm

The first-stage arm is connected with the second-stage arm and the main structure. The connections are removable. During the working period of the whole system, the first-stage arm is a fixed part, the main body is a 20mm thick steel plate whose reinforcement frame is wrapped outside to increase the stiffness of the arm.

4) Second-stage arm

The second-stage arm is connected with the first-stage arm through the hinge structure and provides the running slides of the cantilever system. The structure of second-stage arm is similar to the usually used bracket. In order to further reduce the friction, the guide rail on the second-stage arm adopts the form of air flotation. The second-stage arm and the air flotation guide rail adopt the integrated design. The air flotation slider can move along the air flotation guide rail in a straight line.

5) Pulley system

The pulley system consists of five fixed pulleys and two moving pulleys. The weight on the left side is balanced with the weight of the unloaded object. The weight of the balance is theoretically the same as the mass of the unloaded object (antenna) with the same stroke. During the unloading process, the vertical motion of the antenna will not produce horizontal movement. Besides, the whole pulley system is in complete equilibrium state. When the horizontal external force is applied to the slider, the slider can move freely along the guide rail. The movable pulley provides the connection interface of the unloaded product, and the slider acts as the interface between the cantilever system and the

IOP Conf. Series: Materials Science and Engineering **727** (2020) 012014 doi:10.1088/1757-899X/727/1/012014

second-stage arm. There are two steering pulleys at the front end of the first-stage arm to ensure smooth connection of the tether when the second-stage arm rotates to different angles.

3. Performance analysis of convenient cantilever-unloading bracket for antenna

3.1. Friction resistance analysis

Friction is an important parameter for antenna unloading. The friction of the system is mainly produced in three aspects: the friction moment between the first and second-stage arms when the ball bearings rotate; the linear sliding (air floatation) friction between the slideway of the second arm and the cantilever system; the internal friction of the pulley group (air floatation); the rigid resistance caused by the bending of the wire rope.

(1) Friction resistance moment analysis of arm-to-arm rotating joint

A ball bearing is used to connect the first-stage arm and the second-stage arm to generate the friction moment of rotating motion. The friction moment of the bearing can be calculated by the following formula: $M=\mu \times p \times d/2$

Among them, M: friction moment, mNm

μ :The friction coefficient of deep groove ball bearing is 0.0015.

P: Bearing load, N

D: nominal bore of bearing, mm

When the cantilever position of the load is at the end of the linear guide rail of the second-stage arm, the bearing is subjected to the maximum positive pressure. At this time, the suspension point is 2.1 meters away from the rotating shaft, the total load is 900 N, and the span of the two rotary joints of the rocker arm is 300 mm. Therefore, the maximum positive pressure P is 900*2.1/0.3=6300N. Thus, the friction resistance moment M is 0.34Nm. Because the rocker arm is composed of two upper and lower rotary joints, the total rotary friction is obtained. The resistance moment is 0.68Nm.

(2) Friction Analysis of Linear Slide

This frictional force is composed of two parts. The first part is caused by the pressure of the load and the air-bearing pulley assembly on the slideway. The second part is caused by the horizontal error caused by the bending of the arm bar with the cantilever device, which results in the force distribution along the slideway direction of the load.

The friction coefficient of the linear slide is about 0.0003, and the pressure on the slide is G=900N.

So: $F_1 = 0.0003 \times 900 = 0.27N$

The components of the load along the slideway are as follows:

$F_2 = (1.65-0.45)/1950 \times 900 = 0.55N$

So the total f= 0.55+0.27=0.82N

(3) Internal Friction Analysis of Air Floating Pulley Set

This friction is composed of static pulley and moving pulley. The friction force of the pulley group is proportional to the positive pressure applied on the pulley, and the coefficient of air floatation friction is 0.0003. There are five static pulleys (only four of them actually generate friction) and two dynamic pulleys. The force of each pulley can be easily synthesized.

(4) Rigidity resistance analysis of Kevlar cable

When Kevlar pulls rope around pulley, due to the change of curvature, relative slip occurs between rope strands, resulting in friction and elasticity, which hinder it from adapting to the curvature of pulley rope groove immediately and offset a certain amount outward. Likewise, when the Kevlar pull rope is wrapped around the pulley, it can't immediately restore the straight line state, but generates a deviation value inward, as shown in figure 2. This increases the force arm around the branch and decreases the force arm around the branch. In order to make the pulley rotate, it is necessary to increase the pulling force around the branch which is regarded as rigid resistance.

$G(R+m)=(G+F)(R-n)$

G: Cave pulling force

IOP Conf. Series: Materials Science and Engineering **727** (2020) 012014 doi:10.1088/1757-899X/727/1/012014

F: Stiffness Resistance
K: Stiffness Resistance Coefficient
Among them, the stiffness resistance coefficient K is generally determined by experimental method. Under general conditions, the stiffness resistance coefficient K of Kevlar rope is 0.005.
So in this system, the rigidity resistance of Kevlar rope can be calculated.

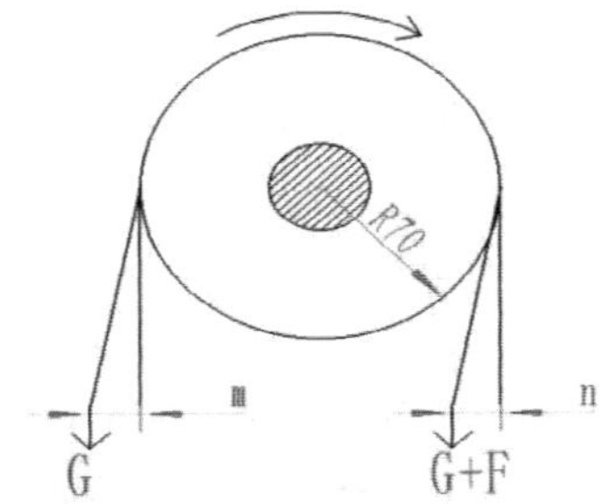

Figure 2. Analysis of Rigidity Resistance of Kevlar Rop

3.2. Anti-overturning design
Because the system itself is a cantilever structure, and the system is under force from upwards when working, it is necessary to carry out anti overturning design for the system, and the position of the system center of mass is shown in figure 3.
According to the system structure, the mechanical model of anti-overturning design is established as follows.

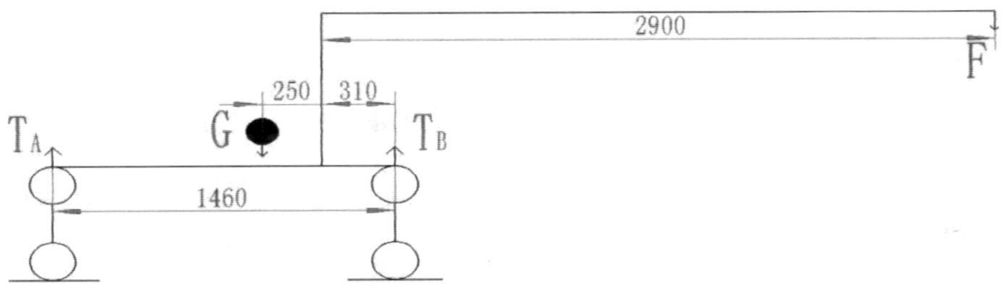

Figure 3. Mechanical sketch of system anti-overturning design

When T_A= 0, the system is analysed: G× (250+310)-F×(2900-310)=0, so, F = 4170 (N).

When F > 4170N, the suspension load > 417Kg, the whole system will overturn around point B. The design load of the cantilever system is 90kg. Therefore, the anti-overturning load capacity margin of the system is 363%, and the anti-overturning margin of the system can meet the application requirements.

3.3. Deformation analysis
The total load of the system is 90Kg. we analyze this condition by assuming the hanging point locating at the end of the second-stage arm, the middle part of the arm and the root of the arm. The angle between the second-stage arm and the first-stage arm is 0 and 90 degrees. The results are shown in table 1.

Table 1. Whole Deformation Analysis Unit of Hanging Device: mm

Position where force loading on the second-stage arm	Maximum Deformation under Rated Load		Maximum Deformation Fluctuation under Maximum Load
	The angle between arms is 0 degree.	The angle between arms is 90 degree.	
at the end of arm	1.656	1.616	0.05
at the middle of arm	1.178	1.024	0.154
at the root of arm	0.711	0.456	0.255

From the table above, it can be seen that the maximum deformation is 1.656 mm under the rated load, which meets the requirement of the maximum deformation of the antenna. In addition, according to the trajectory of antenna, the angle of the second-stage arm relative to the first-stage arm is about zero during the unloading of the antenna with 1700mm working distance. Therefore, the deformation during the unloading process meets the requirements of the index.

4. Summary

In this paper, a universal convenient cantilever-unloading bracket for antenna is proposed, which can be easily moved during deployment/closure test during spacecraft AIT. It not only optimizes the traditional operation mode, but also reduces the operation of the spacecraft and the development time of the spacecraft. Besides, the bracket has high unloading accuracy and suits for many working conditions. It can meet the needs of two-dimensional and three-dimensional unloading conditions and reduce the operational risk. The research results of this paper can be widely used in deployment/closure tests of deployable components of various spacecrafts, such as antennas, solar arrays, etc.

References

[1] Erich Lutz, Steeve Montminy, Zheng Wan ping. Co-channel interference in high-throughput multibeam satellite systems, Communications(ICC)/2015 IEEE International Conference:885-891
[2] HOU Jian, WANG Xudong, etc. Gravity uploading of space deployable overlapping antenna, Space Electronic Technology, 2017,14(6):60～64
[3] LI Tuanjie, MA Xiaofei, Technologies of Large Deployable Space Antennas, Space Electronic Technology,2012,9(3):35～43
[4] LI Yuqi, SHAO Zhufeng, etc. Analysis and Evaluation on Unloading Ratio of Zero-g Simulation Device of Space Manipulator Based on Cantilever system. Robot,2016,38(3):293～300
[5] Han Jiaxu, Kinematic analysis and gravity unloading scheme design of scissors deployable antenna. Yanshan University,2018
[6] HOU Guanzhong, XU Yu, etc. Investigation on Zero-g off-loading in Mechanical Turn Satellite Antenna on Ground. Space Electronic Technology,2018,15(1):74～78

IOP Conf. Series: Materials Science and Engineering **727** (2020) 012015 doi:10.1088/1757-899X/727/1/012015

A Method for Calibrating Thrust of Satellite Based on the Attitude Variety On-orbit

Jialong Ji*, **Yong Gao, Ali Zhang and Qian Lin**

Beijing Institute of Control Engineering, Beijing Engineering Research Center of Efficient and Green Aerospace Propulsion Technology, Haidian District, Beijing 100094, China

* Corresponding author's E-mail: zggdstjl@163.com

Abstract. The propulsion system of satellites is usually used for orbital controlling and attitude-adjusting. And the thrust of thrusters is always calibrated by ground test. A method of calibrating the small thrust of thrusters used for satellites on-orbit is presented in this paper. The application of this method is based on the variety of the attitude of the satellite. And the rotate speed of reaction wheel, the star sensor, and the scopperil, and other components in satellite are utilized in this method. The ground test of 1N thrusters was also carried out, and the thrust calibration test of the same thrusters was performed on-orbit according to this method described here. Both the test were compared, and the results show that both the data are coincident, which means that the method is fit for calibrating small thrust of satellites on-orbit.

1. Foreword

The propulsion system of satellite is widely used for orbital controlling and attitude-adjusting. When the orbital controlling is performed, one or several thrusters work together to increase or to decrease the velocity of satellite, and the centroid of satellite should be in the direction of resultant thrust. It is because that if the centroid of satellite is not in the direction of force, there will be a disturbance torque to satellite, which will consume some more propellant to eliminate the attitude deviation. On the other hand, the attitude-controlling thrusters, which are always located at the corner of satellites and which force doesn't go through the centroid of satellites, are used for adjust the attitude of the satellite. At the same time, Small thrust liquid pulsed rocket engines operating in pulsed mode have gained a good reputation in attitude control applications for their potential reliability and efficiency[1].

Some method is used for thrust test in the laboratory[2], but for the satellite on-orbit, the thrust calibration is always performed while the satellite is carrying out orbital controlling, and the thrust is test by the velocity increment of aircraft. While, if some new propulsion technology which may be used for satellites is validated on-orbit, the thruster may not be located through the centroid of the satellites, and the resultant thrust of thrusters may not go through the centroid either. At this time, the calibration method based on the velocity increment isn't applicative. At the same time, if there is something wrong with the attitude-adjusting thrusters, there should be some diagnosis measures to check the fault. Another method based on the attitude variety on-orbit is available to calibrate the thrust.

IOP Conf. Series: Materials Science and Engineering **727** (2020) 012015 doi:10.1088/1757-899X/727/1/012015

2. General Approach

2.1. System Design

The thrust calibration mainly focuses on the thrust of thrusters, and the thrust lies on the pressure of the system. Take monopropulsion system as an example, a simplified, verified system is shown in figure 1. There are 2 thrusters configured in this system. It is assumed that both thrusters are located as attitude-controlling mechanism, which means that its force doesn't go through the centroid of the satellite.

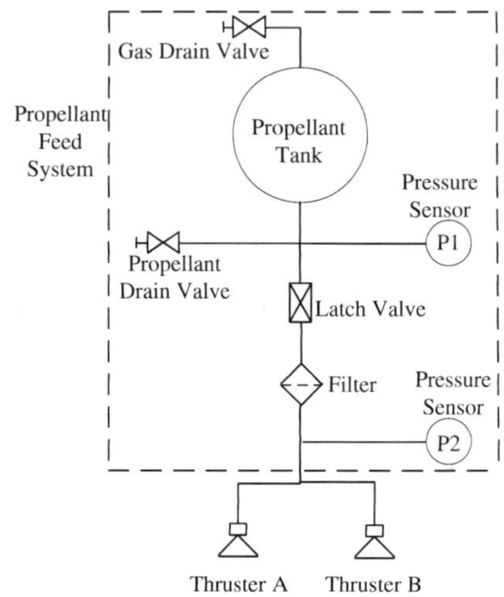

Figure 1. The MonoPropulsion System Sketch

2.2. Calibration Method

By now, the satellite is always designed as a fully 3-axis stabilized craft by the GNC(Guidance, Navigation, Control) system. The thrust calibration is based on the attitude variety of the satellite, so the GNC system of satellite should be applied, and some sensor technology such as reaction wheel(RV), star sensor(SS) and scopperil should be utilized. The demonstrations are described as follows:

a) The thrust calibration is performed by GNC autonomous maneuvering with high precision of autonomy including SS and scopperil measuring the attitude, RV controlling the attitude. And during the calibration, the thrust is validated with the shift of RV's rotate speed, the magnetic torque should not be utilized to unload the rotate speed of RV.
b) Both impulse test and steady test are performed for thrust test and specific impulse test.

The thrust vector to the fully 3-axis stabilized satellite when the thrusters work can be presented as follows:

$$F = \begin{bmatrix} \text{Fx} \\ \text{Fy} \\ \text{Fz} \end{bmatrix} = f \begin{bmatrix} \cos(\alpha) \\ \cos(\beta) \\ \cos(\gamma) \end{bmatrix} \tag{1}$$

The shift of RV's rotate speed can be described as [dn1 dn2 dn3 ...] by the telemetry data of satellite. There are always 4 reaction wheels fixed in the satellite for a complementary function of attitude controlling, and according to the number, the fixed angle, perform ability of the RVs, the shift of angular momentum is:

$$\begin{bmatrix} dHx \\ dHy \\ dHz \end{bmatrix} = \begin{bmatrix} a_{11} & a_{12} & a_{13} & a_{14} \\ a_{21} & a_{22} & a_{23} & a_{24} \\ a_{31} & a_{32} & a_{33} & a_{34} \end{bmatrix} \begin{bmatrix} dn1 \\ dn2 \\ dn3 \\ dn4 \end{bmatrix} * C \tag{2}$$

At the same time, the influence of the thrust to the attitude of satellite is:

$$\begin{aligned} F_z * y - F_y * z &= dH_x/dt \\ F_x * z - F_z * x &= dH_y/dt \\ F_y * x - F_x * y &= dH_z/dt \end{aligned} \tag{3}$$

When calculating the thrust, the vector of the largest angular momentum shift can be utilized. For example, the angular momentum shift of Y axis is the largest, which can be described as follows:

$$f[cos(\alpha) * z - cos(\gamma) * x] = dHy/dt \tag{4}$$

So the thrust of the thruster can be solved by

$$f = \frac{d(\Delta Hy)/d(\Delta t)}{cos(\alpha)*z - cos(\gamma)*x} \tag{5}$$

From the method described in this chapter, we can know that the shift of RV's rotate speed is linear to the thrust. And according to the hot-fire test results of thrusters, the chamber pressure of the first few impulses is usually lower than expected[3]. And after the first few impulses, the thrust and specific impulse is equal to designed. In order to calibrate the performance of steady-state test on-orbit, a method utilizing the linear relation can be applied here, and the test data of the number of impulses, the shift of RV's rotate speed during different stages test will be used:

$$f = \frac{d(\Delta Hy)/d(\Delta t)}{cos(\alpha)*z - cos(\gamma)*x} \tag{6}$$

3. Thrust Calibration

According to the approach presented in the former part, the thrust calibration is performed on-orbit.

3.1. System Design[4]

The propulsion system designed as the system sketch shown in figure 1 is located in a satellite weighting about 4,000kg. This satellite is a fully 3-axis stabilized craft with 4 reaction wheels to adjust the attitude. This propulsion system includes drain valves, propellant tank, latch valve, filter, pressure sensor, and so on. The design object of the system is to generate the moment of force when the thruster works. There are 2 1N monothrusters in the system, and the 2 monothrusters are located in the same surface of the satellite. If one of the thrusters is out of work, another one can still be used to test. In order to reduce the time of unloading the RV's rotate speed, and in order to perform the calibration of both thrusters, the thrusters are located in the symmetrical position, as shown in figure 2.

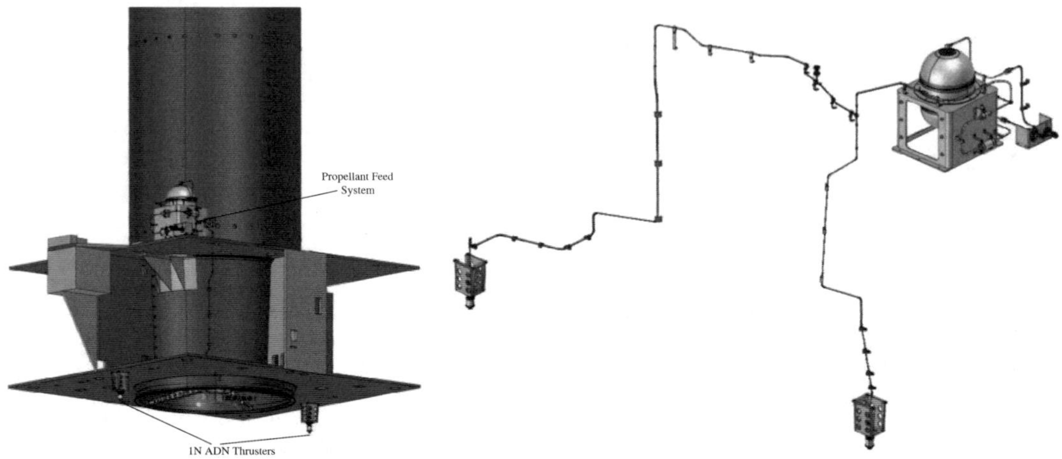

Figure 2. The Propulsion System Layout

3.2. In-Flight Experiment

The propellant was drained into the propellant tank before the satellite was sent to orbit, and the tank was also fed with helium of 1.2MPa. The helium is utilized to push the propellant into the thrusters. The thrusters are designed as the attitude-controlling and orbit-controlling organs. During the thrust calibration, the thrusters are used as the attitude-controlling organs. And during the test, some real-time telemetry data such as the tank pressure, the condition of latch valve, the convolution of RVs, the attitude of the satellite, are obtained from the test and control system(TT&C system).

During the test, impulse test of different impulse number are performed. Among the impulse test, different length of 0.1s/1s*10, 0.1s/1s*20, 0.1s/1s*50 test are performed. The thrust is validated by equation(5) and (6), and the validation is shown in figure 3. From the test, we can know that the thrust ranges from 1.15~1.33N.

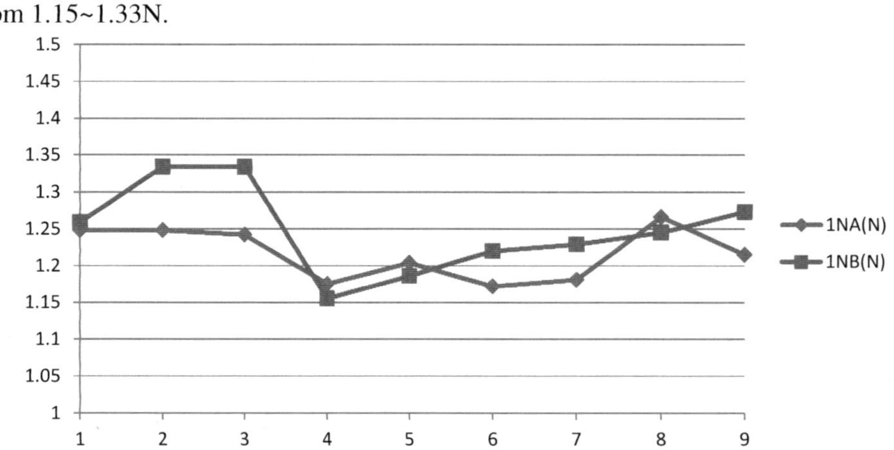

Figure 3. Thrust Calibration of Each Test

3.3. Ground Test

In order to validate the performance of the thrusters, the ground test are performed. Some thrust test of the thruster is carried out, as shown in figure 4. Also, a ground test system using the same components applied in the in-flight system is constructed for hot-firing. The test system is designed as a module, and is put in the vacuum cabin during hot-firing test as shown in figure 5. During the test, the negative step response method is applied[5]. The test result is shown in figure 6. From the test we can know that the thrust is about 1.15N at the pressure of 1.2MPa in the tank.

Figure 4. Steady Test of Thrusters

Figure 5. Hot-firing Test of Propulsion System in Vacuum Cabin

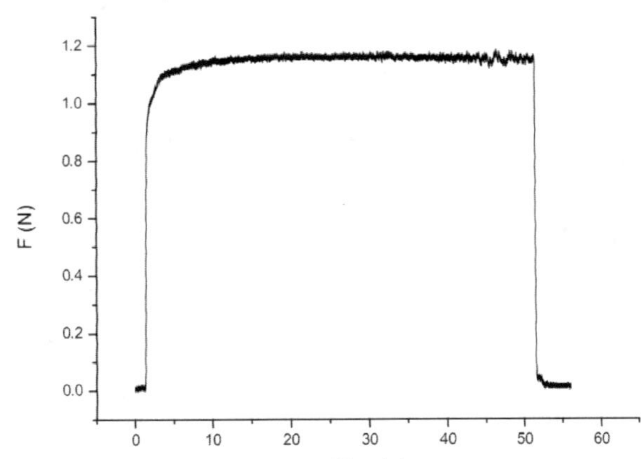

Figure 6. Thrust Test 50s Steady Hot-firing

4. Test Comparison

From the test data above, we can know that the calibration results by in-flight experiment and ground test are consistent. At the same time, the thrust validated inflight is a little higher. The main reason may be the pressure difference between the vacuum cabin and the inflight space. The pressure of the vacuum cabin is higher because of the competence limit of the pump which leads that the cabin pressure cannot be zero, and the gas generated from the propellant increases the pressure of cabin, while the inflight pressure is vacuum. The pressure difference leads to a thrust difference. Another reason may be the devices' measuring precision and the indetermination of satellite's centroid.

5. Conclusions

It is described about the method calibrating thrust on-orbit, the method is based on the attitude shift of satellites. The method is presented here, with some data obtained and telemetered to the ground. Also, this method is validated by calibrating the monothrusters on-orbit. Compared with the ground test result, we can know that the thrust is coincident. The conclusions are achieved as follows:

a) Two thrusters located in this method can reduce the time for preparing the test.
b) This method based on the attitude variety can be utilized to calibrate the small thrust of thrusters in the satellites.
c) There are some differences between the calibrating results of inflight test and ground test, but the test results are consistent.
d) Further research on this method such as the method optimization, the precision increase can be carried out.

References

[1] Qin Xing, Jun Zhang and Min Qian. Design, calibration and error analysis of a piezoelectric thrust dynamometer for small thrust liquid pulsed rocket engines. *Measurement*, 0263-2241, 2011
[2] J K Lam, S C Koay and K H Cheah. Electromagnetic calibration system for sub-micronewton torsional thrust stand. AEROS Conference 2017 Journal of Ordnance Equipment Engineering, Vol.38, No.12, 163-166.
[3] Jun Chen, Guoxiu Li, Tao Zhang and Meng Wang. Research on key components test and ignition performance of green high performance ADN based thruster. 2017 J.
[4] Jialong Ji, Zhonghan Liu and Yuan Wang. Research and in-flight experiment of ADN-propulsion technology for satellite application, GLEX2017, 2017.
[5] Jingsong Gong, Lingyun Hou and Wenhua Zhao. Study on unsteady response characteristics of small thrust measurement system[J](in Chinese). *China Measurement & Test*, Vol.38, No.2, 2012

IOP Conf. Series: Materials Science and Engineering **727** (2020) 012016 doi:10.1088/1757-899X/727/1/012016

Design and Optimization of Vibration Isolation System for Spacecraft Transport Package Box

Guo Tao[1,2,*], Liu Guangtong[1,2], Zhang Jie[1,2], Li Xinyu[1,2] ,Wang Wei[1,2] and Guo Jinghui[1]

[1]Beijing Institute of Spacecraft Environment Engineering, Beijing 100094, China
[2]Beijing Engineering Research Center of the Intelligent Assembly Technology and Equipment for Aerospace Product, Beijing 100094, China

[*]Corresponding author e-mail: guotao102@163.com

Abstract. The damping effect of vibration isolation system for spacecraft transport package box is directly associated with superior or inferior mechanical environment for spacecraft transportation. In this Article, the design process and program for vibration isolation system for spacecraft transport package box were introduced first; followed by analysis on excitation and response during spacecraft transportation by train; finally, an optimization program of vibration isolation system applicable to current working conditions for transportation was given. The results showed that: the frequency of railway road spectrum excitation was concentrated at 1.8Hz and 6.2Hz, indicating that the current vibration isolation system for package box did not have better damping effect. After reducing first-order inherent frequency by about 3Hz, acceleration root mean square may reduce by 40%.

1. Introduction

After research and development, spacecrafts may be delivered to launching site for launching missions by road, train, and ship or air transportation. During transportation, spacecrafts may be subject to excitation caused by disturbance from the environment and means of transportation, including bumps in the road, wave impact, flow disturbance and emergency braking [1-2]. The vibrations and shocks induced by these factors are transmitted to the spacecraft. If not effectively suppressed and attenuated, the spacecraft will be adversely affected or even damaged [3]. Damping is generally achieved by mounting dampers on the base of the package box. For different types of spacecraft, in case of changes on spacecraft mass and center of mass, the number and layout of dampers may be adjusted to minimize the damage caused by vibration during transportation.

From consideration of vibration isolation, the softer the vibration isolation system is, the better the vibration is. But for consideration of the support and the limit, the harder the vibration isolation system is, the better it is to ensure the compactness of the space structure. Obviously these two are contradictory, so in actual application, we should find a compromise solution that can better meet the two requirements [4-5].

This Article introduces the design method of vibration isolation system for spacecraft transport package box, analyzes the data of train running test under true working conditions, evaluates the vibration isolation system of the current transport package box, and finally gives the optimization program for of vibration isolation system.

2. Design of Vibration Isolation System

2.1 Design Process

The vibration isolation system must fully consider the transportation conditions during design, and combine the quality characteristics of the vibration reduction object and the type of damper to finally confirm a reasonable and feasible damping program. For vibration isolation system of spacecraft transport package box, wire rope dampers are generally used. In order to balance vertical and lateral stiffness of the vibration isolation system, wire rope dampers are generally subject to 45° layout [6]. See Figure 1 for design process for vibration isolation system of spacecraft transport package box.

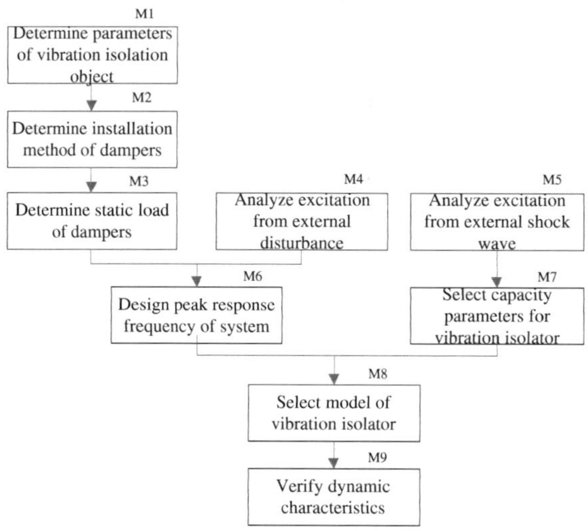

Figure 1. Design Process of vibration isolation system

M1: Determine parameters of vibration reduction object, including weight and dimensions of spacecraft; M2: Determine installation method of dampers: the number of wire rope dampers shall meet requirements on spacecraft stability, installation location shall be distributed evenly, and the stiffness center of the wire rope damper coincides with the whole-spacecraft position in vertical direction as much as possible; M3: Determine static load of dampers; M4: Analyze excitation from external disturbance, and determine peak frequency of external disturbance source; M5: Analyze excitation from external shock wave, shock waves are mainly from working conditions such as emergency braking, turning and hump or dip in road, determine maximum amplitude of the shock wave; M6: Design peak response frequency of system, and determine inherent frequency of vibration isolation system based on excitation from external disturbance, which shall be $1/\sqrt{2}$ less than the frequency of external disturbance; M7: Determine capacity parameters for wire rope dampers based on maximum peak of external shock wave; M8: Select suitable type of wire rope dampers based on number of wire rope dampers, inherent frequency of vibration isolation system and capacity parameters of wire rope dampers; M9: Calculate actual dynamic characteristics (peak response frequency, vibration and shock transmissibility, shock resistance of system) based on actual load-deformation characteristics of wire rope dampers, verify whether meeting with requirements.

2.2 Design Program

Transportation status of model-x spacecraft is shown in Figure 2, the vibration isolation system is supplied with 18 dampers, and L bracket, docking flowerpot and spacecraft above the damper. During transportation, external excitation may be transferred to L bracket through the vibration isolation system before finally transferring to the spacecraft. Controlling the transfer function of vibration isolation system can effectively reduce the transmissibility of vibration power flow.

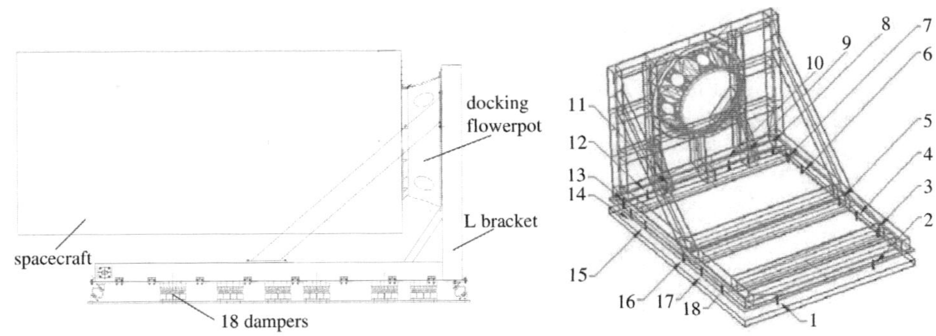

Figure 2. Whole-Spacecraft Transportation Status

3 Analysis of Excitation and Response

3.1 Sensor layout

Train running test under true working conditions may be required to objectively evaluate the efficiency of vibration reduction during spacecraft transportation [7]. During training running test, model-x spacecraft was supplied with 6 vibration sensors and 3 shock sensors as shown in Figure 3. For which, shock 3, vibration 5 and vibration 6 were set at base of package box, the measured data was regarded as road spectrum excitation data of working conditions for transportation. Vibration 2, vibration 3, vibration 4 and shock 2 were set at lower surface of the docking flowerpot. Vibration 1 and shock 1 were set at upper surface of the L bracket, and the measured data was regarded as response data after vibration reduction. Direction X shown in Figure was forward direction of transportation, direction y was lateral direction and direction z was vertical direction.

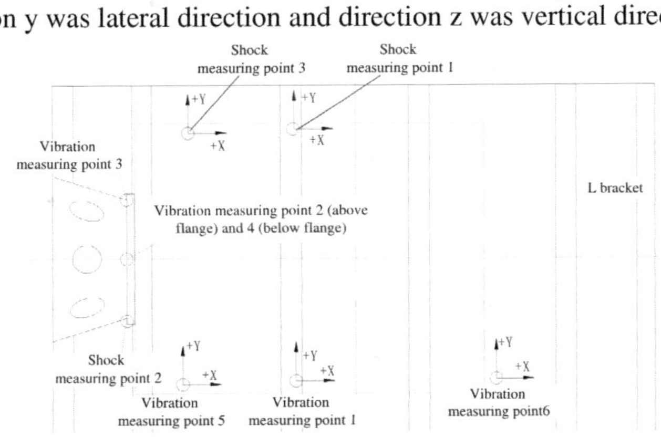

Figure 3. Sensor Layout for Train Running Test

3.2 Analysis of excitation and response

The train running test was carried out according to true path of spacecraft transportation and the typical section on such path was selected for analysis. Figure 4 gives the vibration acceleration power spectral density curve of measuring point 5 and 6. From which we can see that, vertical acceleration power spectral density at vertical direction was far greater than that at forward and lateral direction. The frequency components of the acceleration power spectral density at three directions were centered at 1.8Hz and 6.2Hz.

Figure 5 gives the vibration acceleration power spectral density curve of measuring point 1 and 2. The curve at measuring point 3 and 4 was basically the same with that of measuring point 2. Measuring point 1 was relatively close to base of package box, making vertical vibration greater than other two directions significantly; measuring point 2 was located at docking flowerpot and far away from base of package box, vibration amplitude in the forward direction increased significantly due to pitching angular speed. The frequency components of the acceleration power spectral density at three directions were centered at 1.8Hz, 6.2Hz and 10~11Hz. Based on road spectrum excitation data and measuring point response data, it can be seen that frequency of railway road spectrum excitation were mainly at 1.8Hz and 6.2Hz, while 10~11Hz shall be a first-order frequency of the vibration isolation system.

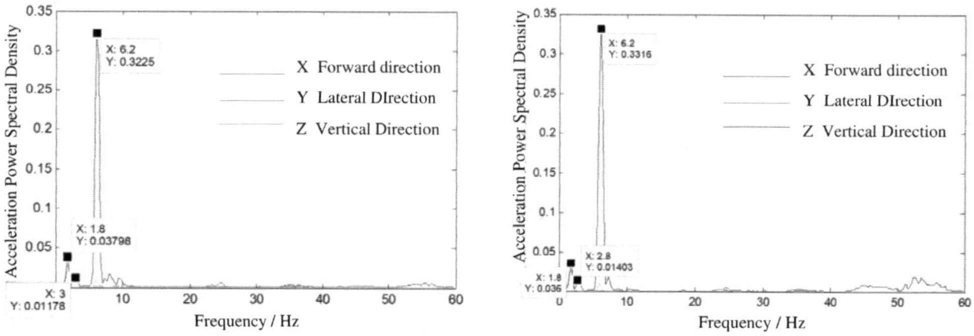

Figure 4. Acceleration Power Spectral Density of Vibration Measuring Point 5 (Left) and 6 (Right)

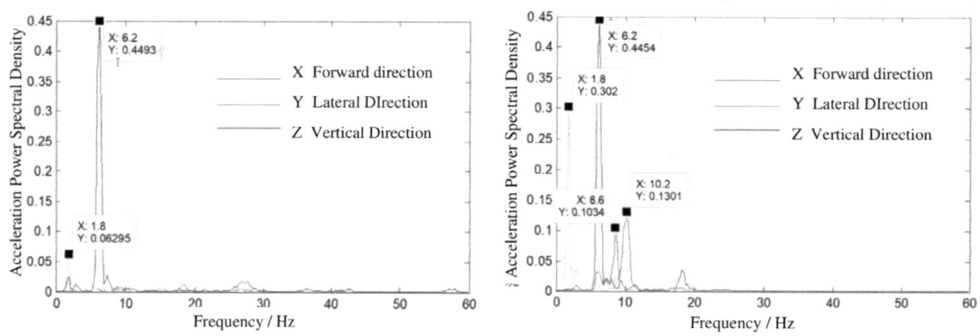

Figure 5. Acceleration Power Spectral Density of Vibration Measuring Point 1 (Left) and 2 (Right)

Figure 6 gives the acceleration power spectral density curve of measuring point 1 and 2 of shock sensor, frequency trend and components of which were basically consistent with the measured data on vibration sensor.

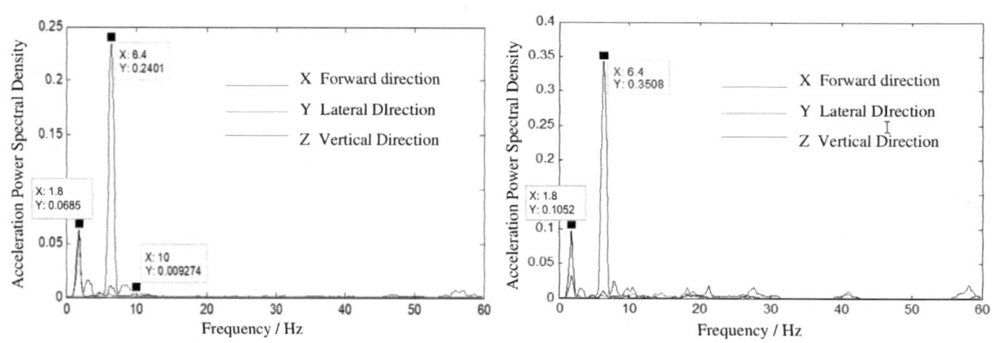

Figure 6. Acceleration Power Spectral Density of Shock Measuring Point 3 (Left) and 1 (Right)

Acceleration root mean square of each measuring point may be obtained by integration processing of acceleration power spectral density curve, as shown in Table 1 and Table 2. For measuring point at vibration sensors, data was amplified averagely by about 2.5times at direction x, about 1.5 times at direction and slightly decreased at direction z, making integrated acceleration amplified averagely by about 1.2 times. For measuring point at shock sensors, data was amplified averagely by about 2.2times at direction x, about 1.53 times at direction and slightly decreased by about 1/2 at direction z, making integrated acceleration consistent substantially. Hence, current vibration isolation system did not have better damping effect and required optimization.

Table 1. Vibration Acceleration Root Mean Square by Measuring Point

RMS	Excitation Point (m/s^2)			Response Point (m/s^2)				
	Measuring Point 5	Measuring Point 6	Mean Value	Measuring Point 1	Measuring Point 2	Measuring Point 3	Measuring Point 4	Mean Value
x-forward direction	0.2027	0.2068	0.2048	0.2802	0.5499	0.667	0.6647	0.5405
y-lateral direction	0.2236	0.2063	0.2150	0.3134	0.5034	0.4107	0.3443	0.3930
z-vertical direction	0.6423	0.7181	0.6802	0.7231	0.6853	0.3908	0.2796	0.5197
Integrated	0.7097	0.7752	0.7425	0.8364	1.0127	0.8753	0.7991	0.8809

Table 2. Shock Acceleration Root Mean Square by Measuring Point

RMS	Excitation Point (m/s^2)	Response Point (m/s^2)		
	Measuring Point 3	Measuring Point 1	Measuring Point 2	Mean Value
x-forward direction	0.2075	0.3096	0.6565	0.4831
y-lateral direction	0.3426	0.3789	0.5396	0.4593
z-vertical direction	0.9987	0.7145	0.4173	0.5659
Integrated	1.076	0.866	0.9467	0.9064

4 Optimization Program

Without changing the overall program of the vibration isolation system, adjusting the number of dampers is the most simple and easy solution for optimizing the vibration isolation system. Table 3 gives the strategy for stiffness of the vibration isolation system adjustment. After adjusting the number of dampers, the positions of other dampers will be adjusted accordingly so that the bearing capacity of each damper is uniform.

Table 3. Stiffness Adjustment Strategy

No.		Adjustment Method	Adjusted First-Order Frequency (Hz)
Increase system stiffness	Strategy 1	Increase 4 dampers	12.08
	Strategy 2	Increase 12 dampers	16.37
Decrease system stiffness	Strategy 3	Remove dampers 3, 6, 15 and 18	6.67
	Strategy 4	Remove dampers 1, 2, 4, 5, 9, 10, 11, 12, 16, 17	3.69
	Strategy 5	Remove dampers 1, 2, 4, 5, 6, 9, 10, 11, 12, 15, 16, 17	3.09
	Strategy 6	Remove dampers 1, 2, 4, 5, 6, 7, 9, 10, 11, 12, 14, 15, 16, 17	2.21

For simulation calculation, the measured data curve of the acceleration power spectrum is generally used as the stimulus input for simulation [8]. That is, the measured data of measuring point 5 and 6 in train running test are used as input for different adjustment strategies, and the vibration responses at measuring point 1, 2, 3, and 4 in the train running test are obtained. Figure 7 gives the acceleration power spectral density at measurement point 2. Table 4 gives the acceleration root mean square of each strategy.

We can see from Figure 7 and Table 4 that, Strategies 1-3 had no significant impact on acceleration power spectral density; acceleration root mean square of strategy 4 decreased by -20.99%, with peak amplified at frequency of 2Hz and reduced significantly at frequency of about 4.4Hz. Acceleration root mean square of strategy 6 decreased by -44.30%, with peak amplified at frequency of 2Hz and reduced more significantly at frequency of about 4.4Hz, indicating better damping effect.

Without changing the model of dampers, strategy 5 was recommended, for which damper deformation was 13mm and meeting capacity requirements. Acceleration root mean square may be reduced by 31.28% within scope of damper application. For strategy 6, though damping effect was more significant, the deformation was 30mm and above, which was close to capacity limit of damper, thus such strategy was not recommended.

Strategy 5 reduced the inherent frequency of vibration isolation system by about 3Hz, which may lead to greater displacement of spacecraft easily in case of large shock. And once the maximum displacement exceeded the capacity limit of damper, spacecraft may be subject to secondary shock. Hence, after implementation of strategy 5, limit blocks and retarders may be recommended on the vibration isolation system to prevent secondary shock to spacecraft in special circumstances. The layout after adding of limit blocks and retarders is shown in Figure 8. Damping of hydraulic retarder may be adjusted as required. And if vertical damping coefficient was adjusted to as twice the previous value, the acceleration root mean square may be reduced by 50% and above as compared with the previous program.

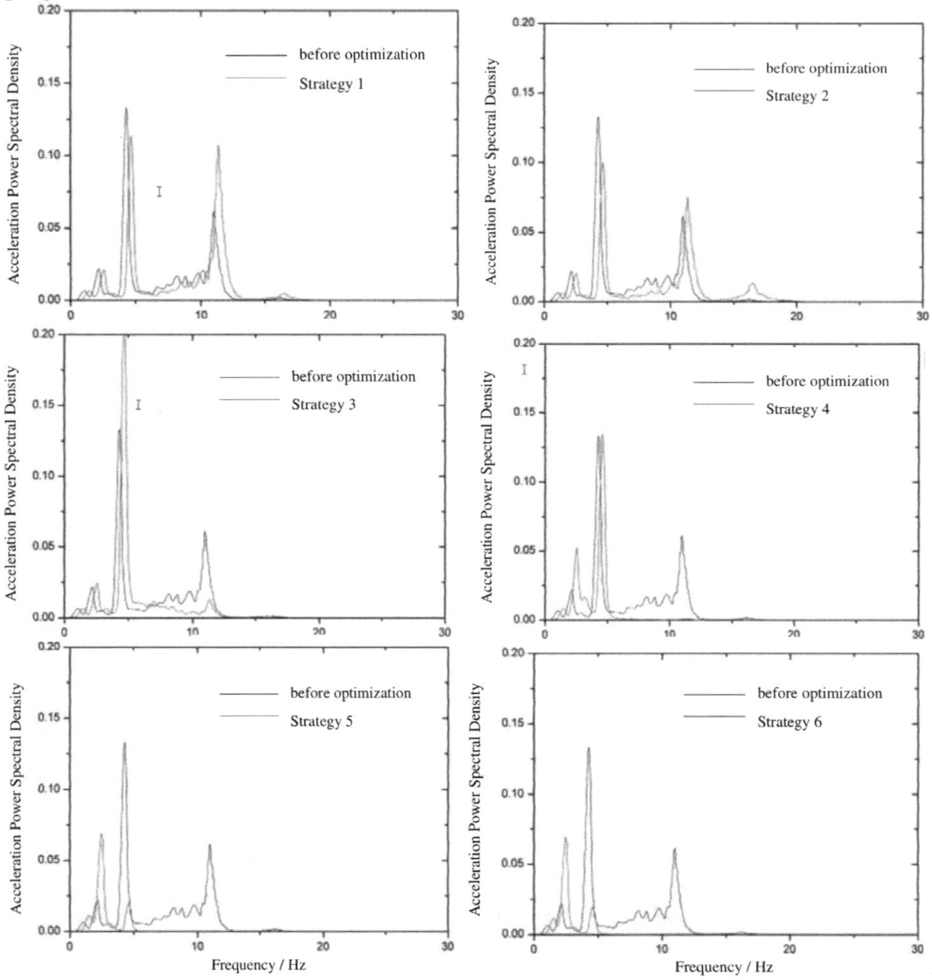

Figure 7. Acceleration Power Spectral Density at Measuring Point 2 in Different Dampers Number

Table 4. Acceleration Root Mean Square at Measuring Point 2

No.		RMS before Adjustment (m/s^2)	RMS after Adjustment (m/s^2)	Variation
Increase system stiffness	Strategy 1	0.4393	0.4709	-7.19%
	Strategy 2		0.4369	0.55%
Decrease system stiffness	Strategy 3	0.4393	0.4244	-3.39%
	Strategy 4		0.3471	-20.99%
	Strategy 5		0.3019	-31.28%
	Strategy 6		0.2447	-44.30%

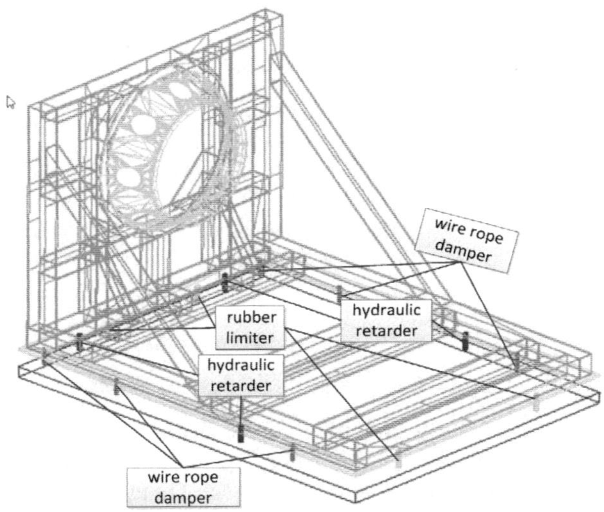

Figure 8. Recommended Layout of Limit Block and Retarder

5 Conclusions

Following conclusions were obtained in this Article after calculation and analysis:

1) Frequency of railway road spectrum excitation was mainly concentrated at 1.8Hz and 6.2Hz;
2) Damping system with first-order frequency of 10~11Hz did not have better damping effect;
3) Damping effect was obvious and acceleration root mean square may be reduced by 40% when inherent frequency of damping system was reduced by about 3Hz;
4) For design of damping system with low frequency, limit blocks and retarders are recommended to prevent secondary shock.

References

[1] Shi Lixia, Luo Ke, Dou Renchao, Feng Qi. Dynamics Environment Excitations During Spacecraft transportations [J]. Spacecraft Environment Engineering, 2013, 30(3): 250-255.

[2] Duan Huming, Shi Feng, Xie Fei, Zhang Kaibin. Summary of research on road roughness[J]. Journal of Vibration and Shock, 2009, 28(9):95-101.

[3] Zhao Ling, Tang Lan, Xiang Chunsheng. Analysis of Damping Performance for Vibration Isolation System of Precision Instrument Transport Vehicles[J]. Journal of Xihua University (Natural Science), 2011, 30(4): 1-6.

[4] Bai Qiaofeng, Liang Hailong, Mei Xiaolong, Liu Guozhi. Simulation Analysis of Suspension System of Spacecraft Transport Packing Containe [J]. Mechanical Engineering&Automation,2013, (2): 19-23.

[5] Zhu Shijian, Lou Jingjun, He Qiwei, Wen Xuetao. Vibration Theory and Vibration Isolation [M]. Beijing: National Defence Industry Press, 2016.

[6] Xiao Gang, Hao Wenyu, Zhang Guofen, Cong Fei. The acquisition, storage and analysis of mechanical data in spacecraft transportation [J]. Spacecraft Environment Engineering, 2010, 27(6): 795-799.

[7] Huang Huaide, Li Xianshan, Bao Daoming. Vibration Engineering [M]. Beijing: China Aerospace Publishing House, 2005.

[8] Kang Tian, Li Chunli, Jiang Huabing. Adjustment method for random vibration test conditions in simulations of road transportation [J]. Spacecraft Environment Engineering, 2015, 32(5): 532-536.

IOP Conf. Series: Materials Science and Engineering **727** (2020) 012017 doi:10.1088/1757-899X/727/1/012017

Application of Reverse Engineering on Spacecraft Assembly Process Simulation

Xu Yiliu[1,2,*], Yang Chunsheng[1,2], Xu Kai[1,2], Wang Wei[1,2],Fan Xing[1] and Hou Jun[1]

[1]Beijing Institute of Spacecraft Environment Engineering, Beijing 100094, China
[2]Beijing Engineering Research Center of the Intelligent Assembly Technology and Equipment for Aerospace Product, Beijing 100094, China

[*]Corresponding author e-mail: xuyiliu0106@163.com

Abstract. This paper proposes a new process that assembly simulation with 3-D model which obtained from product to improve the reliability of spacecraft assembly. 3-D model of the satellite camera which obtained by reverse engineering was used in DELMIA virtual assembly simulation. The point cloud date of the model was scanned by laser radar, and the scanning accuracy is 0.3mm. To get more accurate model data, firstly it's necessary to remove noise data by the method of HCI and curve inspect; then, min-distance method was used to streamline the point data, and repair the data by complementary surface; finally, reconstruct the model by the triangular Bezier surface. Because the model obtained by RE is more exact than the ordinary model, the virtual assembly simulation is more useful .During the period of installing the camera, as the simulation, the risk was avoided and the process was successful.

1. Introduction

Spacecraft products are featured by high functional density, high sensitivity, and complex shape. In order to avoid product interference during assembly, to ensure product safety and ensure the perfect completion of assembly, it is required to identify the risks in assembly in advance during the stage of process design and preventive measures should be developed accordingly.

Virtual assembly simulation is an effective way to identify the risks in complex assembly processes, and the validity of the simulation results depends largely on the accuracy of the product model. In a real case, the outer surface of some product may be a complex free-form surface, which is difficult to be described by a strict and unified mathematical language. The design model of a product often cannot accurately reflect the appearance parameters of the real product. For example, the details such as instrument angle and electrical connectors and flexible entities such as thermal control multilayers cannot be fully described in the design model. Therefore, the real assembly process cannot be accurately simulated using existing design models.

The physical 3D model reconstructed by reverse engineering can more accurately reflect the complex surfaces and flexible entities that cannot be accurately described in the traditional 3D model, and the reconstructed model has a high similarity with the real one. The assembly simulation using the accurate 3D model obtained by reverse engineering can effectively verify the feasibility of assembly and detect the interference points and risks in advance.

This research focuses on the virtual assembly simulation of a certain type of camera on a satellite. The reverse engineering technology is used to reconstruct the camera model. The reconstructed model is used to simulate the assembly process. The risks and challenges in the assembly process are successfully identified, and the countermeasures are given. Finally, this type of camera is successfully assembled, showing the great potentials of the application of reverse engineering to the assembly process design of spacecraft.

2. Overall design

The overall process of virtual assembly simulation is shown in Figure 1. The data scanning is to obtain the point cloud data of the measured object by means of measurement. The model reconstruction in reverse engineering is to extract the points, lines and surfaces required for the reconstruction through the measurement and analysis of the object, and to construct the 3D model by surface reconstruction technique [1]. Digital measurement technology is the basis of reverse engineering and the main means of digitizing the surface of the object. Choosing the right measuring instruments and testing methods to obtain the high-precision features of the object surface is the basis and key to achieving the reverse modeling. In this research, laser radar is used to perform 3D scanning on the physical object to obtain its point cloud data.

The point cloud data obtained by 3D scanning cannot be directly used for model reconstruction due to the great noise, massive amount of data, and "holes" in scanning process. Therefore, it is necessary to perform data preprocessing, to remove the noise, streamline the data, and repair the "holes" before model reconstruction. The inverse reconstruction is to reconstruct the object surface using the processed point cloud data, and finally generate a 3D model of the object. The accuracy of the reconstructed model is closely related to the techniques used in data preprocessing and surface reconstruction.

After obtaining the accurate 3D model of the object, we can start to simulate the virtual assembly process, analyze and verify the assembly path, the feasibility of the assembly, identify the operation risks, and develop the countermeasures, thereby improving the reliability of assembly process.

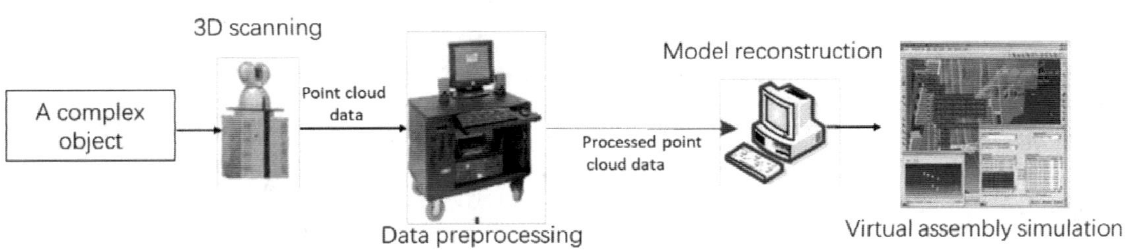

Figure 1. Overall process of virtual assembly simulation

In this paper, a camera of a certain type is used as the research object. The camera has a complex outline, a large number of protrusions, and a large area of the camera surface is covered with thermal control multilayers. Thus, the design model cannot truly reflect the physical shape. The camera is installed in a relatively small and enclosed cabin. Therefore, the safety distance between the camera and the cabin is small, which makes it prone to have bump or interference. As a result, it is necessary to pre-identify the assembly risks through virtual assembly simulation.

The cabin where the camera is mounted has a rigid structure. After the 3D scanning on the cabin, the result shows that the scanned model is the same as the design model, and can completely reflect the actual state of the cabin to be assembled. The camera model for assembly simulation is obtained through reverse engineering. The cabin to be assembled adopts the existing design model.

3. 3D scanning

Data measurement in reverse engineering refers to measure the outline information of an object through a measuring device. The measurement methods include contact measurement and non-contact measurement [2]. In this research, the outer surface of the camera is covered with multi-layer thermal insulation components, and the component surface is made of a flexible material. Therefore, non-contact measurement is adopted in this research.

IOP Conf. Series: Materials Science and Engineering **727** (2020) 012017 doi:10.1088/1757-899X/727/1/012017

Laser radar is used for 3D scanning. The virtual vision scanning mode of the laser radar can collect a large amount of data during the process of scanning a certain surface. The laser radar model is MV330, with scanning precision of0.3 mm in the virtual vision mode, scanning speed of 1000 points per second. The steps for performing laser radar-based 3D scanning are described in the following subsections.

3.1 Divide the scanning area and determine the reference points

As the camera surface coated with the multi-layer thermal insulation components is a free-form surface, the data is collected according to the surface features and the flow direction in order to collect the effective data. The measurement area is usually divided into several sub-areas according to the distribution of the surface features, and the division generally follows the following principles [3]:

a). Scan the entire surface as much as possible. If the surface cannot be scanned with one time, minimize the number of sub-areas as much as possible.

b). The division of the measurement area should avoid the boundary line from crossing or intersecting the feature line. It is better to divide the transition area into one measurement sub-area to ensure the integrity of the feature data.

c). The direction of the feature lines in the measurement area should be as uniform as possible.

d). To ensure the integrity of the data, there should be some overlap in the measurement area.

According to the principle of surface division and the surface features of the camera, the scanning area is divided into five sub-areas, and the 3D scanning is performed on the five sub-areas. In order to establish the relative coordinate relationship of each sub-area, 14 balls are used as the reference points on the camera to establish reference point coordinate system that covers different angles of view.

3.2 Setting of sampling density

In the process of 3D scanning, different sampling densities can be set for the surfaces of different features, in order to improve the scanning efficiency with accuracy maintained. After analyzing the surface features of the camera, in order to more realistically reconstruct the contour, a small scanning density of $0.3\ mm \times 0.3\ mm$ is set for the irregular surface covered with multi-layer thermal insulation components and the mounting surface between the camera and the cabin. By contrast, for a more regular surface, the scanning density can be appropriately enlarged to $0.5\ mm \times 1.0\ mm$ without involving the feature boundary and the critical dimension. During the scanning process, the scanning density of some boundary areas will be higher than the actual scanning density due to the overlap of scanning area.

3.3 Data measurement in virtual vision mode

When using the virtual vision scanning mode of the laser radar for reverse engineering, the first step is to determine the boundary of the scanning area according to the divided scanning sub-areas. The contour points (i.e. the boundary points of scanning area) are measured in the SAMVx software. A certain number of contour points are scanned according to the spatial order and area features, and then these points are connected to form an enclosed curve as the scanning area of the current scan.

After the contour setting of each scanning area, the laser radar can automatically perform data scanning and acquisition in the area. There is a need to ensure that the data of each scan is in the same coordinate system. So, after the coordinate system is established in the first scan, the reference points of balls are first transferred before each scan. In order to reduce the error caused by the transfer, at least 6 points are selected for measurement at the time of the transfer. Then the uncertainty analysis is performed to remove the points with large error, and the error of the transfer is finally controlled within 0.01 mm. The data scanned at different angles is displayed in one coordinate, as shown in Figure 2.

Figure 2. Point cloud data after coordinate fitting

4. Data preprocessing

Data preprocessing is an important step of reverse engineering as it determines whether the subsequent reconstruction process can be carried out in a convenient and accurate manner. In order to better complete the model reconstruction, the raw measurement data must be preprocessed.

1) Removal of noise

The existence of great noise directly leads to a large difference between the reconstructed model and the physical object. Thus, it is necessary to identify and remove the noise in the measurement data. In addition, in order to improve the modeling speed of the subsequent reverse model, the redundant data should be removed as much as possible while ensuring the accuracy of the measured surface. Commonly-used noise filtering methods include human-computer interaction method, curve inspection method and string height difference method [2].

In this paper, the human-computer interaction method and the curve inspection method are combined to remove the noise of the curved point cloud of the camera. Firstly, the human-computer interaction method is used to remove the obvious useless points in the point cloud image. Then, the curve check method is used to discriminate the point cloud by the fitted data curve, and the noise is identified and removed.

2) Streamlining of data

In the reverse operation process, too much point cloud data not only reduces the processing efficiency, but also has a serious impact on the smoothness of the surface. Therefore, it is necessary to streamline the measured massive data. Commonly-used methods for data streaming include: minimum distance method, angle deviation method, uniform network method and adaptive minimum distance method [1]. The adaptive minimum distance method calculates the scanning curvature based on the minimum distance method, and sets different angle thresholds according to the curvature change, thereby ensuring that the scanning line data retains considerable quantity of points in the place where the curvature changes greatly, and retains relatively few points in the place where the curvature changes gently. So, we can keep the micro-features of sudden-change places while streamlining the data, which is a good method in terms of accuracy, simplicity and speed [4]. The algorithm of curvature-based sampling in the surface reconstruction module Pro/SCANTOOLS of Pro/E software adopts this algorithm. This paper uses this module to streamline the point cloud data.

3) Repair of data points

Currently, the widely-used methods in reverse engineering include modeling design method, curves and physical filling method, and surface interpolation method, etc. [5] In order to obtain a complete 3D model that is basically consistent with the physical contour, the supplementary measurements are performed on the undetectable areas with obvious surface features to ensure the integrity of the data. As for the small holes still exist in the later stage, they are mainly repaired by the surface interpolation method. In the filling and repairing process, the curvature of the circled points is calculated, and the segment of point data is filled based on the curvature calculation result. Therefore, we can ensure the surface features in the filling area, and ensure that the repaired small holes are as close as possible to the real object. Hole repair is achieved by the tool in Pro/E, and the data to be repaired is repaired after estimating the curvature of the circled points. Figure 3 shows the point cloud data after preprocessing.

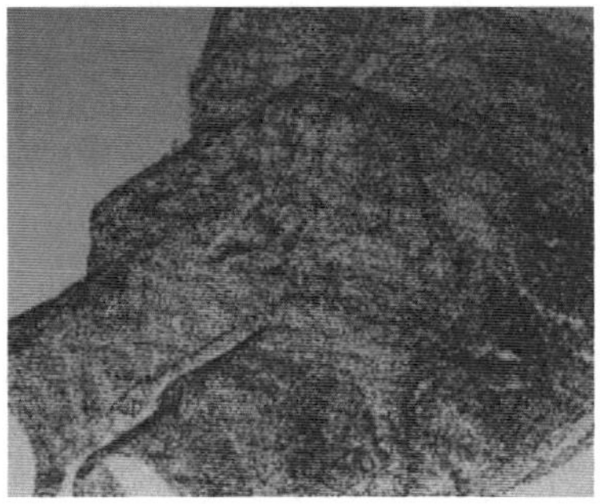

Figure 3. Point cloud data after coordinate fitting

5. Reverse reconstruction of the model

Surface reconstruction can be achieved by two methods. One is to represent the model using NURBS surface, where the parameters are automatically segmented according to the specified surface range, and then the control points of the NURBS surface are calculated using the point cloud data. The other is to represent the model using triangular Bezier surface, and the specific steps are as follows. Firstly, any three points of the point cloud data that form a number of facets are interpolated and reconstructed using the triangular Bezier surface. As the calculation of NURBS surface consumes large resources, and the noise of data as well as the effect of parameter segmentation has great influence on the accuracy and smoothness of the reconstructed surface [2], this paper chooses the method of triangular Bezier surface.

The triangular Bezier surface method is most suitable for representing the objects with irregular complex surfaces. Firstly, the measured point cloud data is divided into several triangular planes, and then the divided triangular mesh is fitted into a Bezier surface. In the Pro/E facet module, the point cloud data is enveloped into a number of small triangular planes by the facet envelope process. After the coarse and fine adjustment of the small triangle planes, the surface patches are defined, and the system will make the small surfaces be reasonably distributed on the surface of the model according to the curvature and composition features of the model. The surface is automatically generated by the command "Create Surface" in the Pro/E menu. After the preliminary surface is created, the curve shape can be adjusted by modifying the mathematical properties and control points of the surface. Finally, through the reconstruction of the surface, such as deleting the small surfaces, reconstructing the boundary lines, etc., the surface is reconstructed, and finally a 3D solid model is generated, with its local view shown in Figure 4.

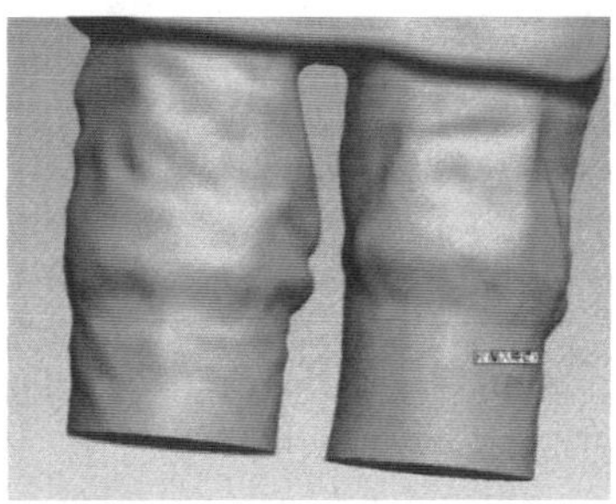

Figure 4. Local view of the reconstructed model

6. Virtual assembly simulation

Assembly simulation is an effective means to verify the detachability/assemblability of a product. The movement pattern and spatial relationship of the components during the assembly process are visualized by relevant visualization methods and interference inspection tools. Besides, assembly path and scheme can also be provided by the visualization. The other aspect is process simulation. The assembly environment is constructed using virtual reality technology, to verify the feasibility of the assembly process, and visually simulate the assembly process in a way close to real assembly [6]. Through three-digit simulation verification, it is possible to find possible challenges and risks from the virtual assembly operation, thereby improving the assembly process and giving corresponding countermeasures [7, 8]. In this paper, DELMIA software is used to realize virtual assembly simulation. The virtual assembly technology based on DELMIA introduces ergonomics factors, which can not only simulate the real assembly process, but also deeply analyze the operability, visibility and accessibility [9].

The satellite model and the camera model reconstructed by reverse engineering are introduced into DELMIA for virtual assembly simulation. The interference detection and collision check in the assembly process are carried out through the collision detection software in DELMIA. The assembly simulation is shown in Figure 5. The risks and challenges identified during assembly simulation shown in Table 1 are consistent with the counterparts found in the real assembly process. The simulated assembly process and the effects of using it in the real assembly process are also shown in Table 1. The developed process effectively avoids the risks in the real assembly, reduces the assembly difficulty, and ensures the smooth completion of the assembly process.

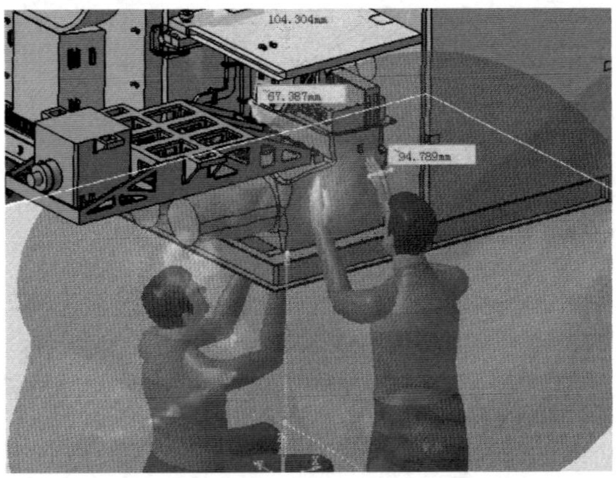

Figure 5. Schematic diagram of the assembly simulation process

Table 1. Comparative analysis of simulation result

	Risks and challenges identified in simulation	Formulated process measures	Analysis of real assembly
1	The camera has a small waveguide spacing on the -Y vertical baffle, which is easy to bump waveguide.	Wrap about 30 mm-thick antistatic foam to the waveguide of the vertical baffle before assembly to prevent damage to the waveguide.	Anti-static foam effectively prevents the camera from colliding with the waveguide and protects the waveguide on the vertical baffle.
2	No tightening and force measurement space for the fasteners between the camera and the bottom frame.	Design a torque adapter rod that meets the current condition, and calibrate the rod in advance.	The newly-developed torque adapter rod successfully solves the force measurement problem under this working condition.
3	The pin on the -Z+Y side is less visible and difficult to align and mount.	Insert the long screw ahead of time near the pin position where fasteners are installed, assist the pin to align, and provide special lighting tools for better lighting.	The equipped lighting tool effectively enhances the visibility at the position of the pin, and the auxiliary screw effectively assists the pin to smoothly align.

IOP Conf. Series: Materials Science and Engineering **727** (2020) 012017 doi:10.1088/1757-899X/727/1/012017

References

[1] Zhang Nian. Virtual assembly simulation and optimization of key parts of steam turbine based on reverse engineering, 2016.
[2] Li Congbo, Gu Xiaojin, Li Linging, Xiao Weihong.Key technologies and applications of reverse engineering aided used part remanufacturing[J],Computer Integrated Manufaturing Systems, 2015, 21(1):143-150.
[3] Meng Yu.Registration of point clouds basing on sample-sphere and ICP algorithm [D]. Beijing: Tsinghua Uniersity,2012.
[4] Chen Y, Yue L,A method for dynamic simplification of massive point cloud[C].IEEE International Conference on Industrial Technology (ICIT). 2016
[5] Zhang Fubao,Huang Mingyu.Rapid design and manufactuer of mold for complicated curve surface part cased on RE.[J].Machinery design and manufacture,2011(7),53-54
[6] Sun Gang,Xiong Tao,Xia Jixia,Zhou Xin,Sun Tao.A vision of using virtual assemble technology in spacecraft assembly[J],Spacecraft environment engineering,2006,23(3),158-161.
[7] Huang Lei,Liu Mengzhou,Xing Shuai,Wei Yue'e.Application of virtual simulation technology in assembly process design of maned spacecraft[J]. Spacecraft environment engineering,2011,28(6),609-614.
[8] Lou Yizhi, Zhou Yongping .Shen Liang,Fei Wanghua.Research on application of virtual assembly technology in aircraft structure design[J],Mechanical research an application,2014(4),125-127
[9] Feng Wei,Zhang Yanlei,Yi Wangmin,Zheng Shengyu,He Wenxing,Feng Baohua,The virtual simulation technology in assembly process design of China's Lnar Exploration ProjectII. Spacecraft environment engineering,2014, (3),326-331

The Test and Verification Methods for the Transportation Vibration Absorption System of Spacecraft

Liu Guangtong*, Xukai, Guotao, Zhang Qiang and Liuzhe

104#You Yi Road Haidian District Beijing China, Beijing Institute Of Spacecraft Environment Engineering, China Academy Of Space Technology

*liugt2007@qq.com

Abstract. The complicated road conditions for the transportation of spacecraft give rise to the need for plenty of running tests so as to validate the performance of the vibration absorption System. According to the principles of subscale test, a research is conducted on the approaches to testing the vibration absorption System for the transportation vibration absorption system of spacecraft. Besides, the subscale model and verification system are designed for the vibration absorption System, and the reliability of the vibration absorption system in transit is validated, which provides a means to conduct effective evaluation of the system for energy efficiency and an approach to the optimization design of it.

1. Introduction

As the aerospace undertakings are undergoing a fast-paced development in China, the number of space missions is on the increase exponentially. Due to the fact that the manufacturing bases of satellites and their launch sites are distributed across different regions, the transportation of aircraft has become a crucial link of type development. Whether the aircraft is transported on highway or by rail, the package used to pack those sophisticated aerospace products would be subjected to various vibrations, which has a potential to affect the sophisticated aerospace products for their performance.

In general, the vehicles used to transport aircraft are complete with a multi-tiered vibration absorption System. For example, tyres, suspension system, as well as the first-tier vibration absorption System and second-tier spring system of rail carriages. Apart from that, the package in itself comes with an independent vibration absorption System. Before transport, the configuration of the vibration absorption System has to be performed based on the characteristics of what is to be transported and the conditions of transport. Nevertheless, the variations of practical driving conditions makes it easy for the vibration data to exceed the specified range when the aircraft is in transit.

The complex conditions of transport lead to an increasingly high requirement on the reliability of aerospace products, as a result of which the development of a testing and verification system as the critical facility to test reliability is increasingly significant. Therefore, it is considered as a critical issue facing the transport of sophisticated aerospace products to construct a subscale model for the package of aerospace products before transport and the vibration absorption System and validate the in-transit reliability of the vibration absorption System.

2. Introduction of the commonly-used methods of testing

Whether at home or abroad, there are two major ways to perform vibration test on transport packaging. One is to load the package onto the vehicle for running test to be conducted in line with the

IOP Conf. Series: Materials Science and Engineering **727** (2020) 012018 doi:10.1088/1757-899X/727/1/012018

specifications on road conditions, vehicle type, vehicle state, driving speed and mileage. The other one is to perform an in-lab simulation experiment to the standard drawn up according to the specific conditions of transport [1,2].

- On-board test

The item under tested is loaded onto a vehicle, which is required to drive a certain distance based on the specified load, speed, as well as road and meteorological conditions. Then, the item is unpacked to check whether it is intact. The merits include simplicity in operation and intuitiveness of test results. On the flip side, the drawbacks include expensiveness, long duration, as well as the difficulties in conducting quantitative analysis and exercising control over the testing conditions.

- Indoor simulation test

This is premised on a thorough study conducted of the vibration data on the practical conditions of transport. It shows such merits as low cost of experimentation, quickness, typical experimental conditions, and easy control. Moreover, it exhibits an excellent reproducibility, which is beneficial to reduce the time and monetary costs of experimentation. On the flip side, it shows the drawbacks like high initial cost of investment and demanding requirements technically.

In vibration test, the test piece needs to be excited based on the actual environment of vibration. According to the nature of the applied load, the test is classed into three different types. The first one is sinusoidal vibration test, where the test piece is subject to sinusoidal vibration, including sine FM and swept-sine vibration tests. The second one is randomized vibration test, where the test piece is subject to random vibration. Based on the size of bandwidth, it can be further classified into broad-band random vibration test and narrow-band random vibration test. The third one is mixed vibration test, where the test piece is subject to both sinusoidal vibration and random vibration simultaneously [3,4]. In transit, the occurrence of vibration is a randomized process. Due to various factors like the unevenness of road surface, vehicle type, loading, driving speed and so on, the vehicle is subject to randomized excitation in transit. Therefore, the randomized vibration test is usually performed for indoor simulation.

3. The design of subscale model

The subscale model is chosen to perform test because some of the aerospace products are characterized by heavy weight and large volume, which means that it would be costly to use the original items for testing. Subscale test is known as a commonly-used approach to verification in engineering. The key to its success lies in the design of a specific subscale model. In principle, the main experimental data obtained from the model and the prototype system must be close. Therefore, the verification test shall start with the determination of subscale model based on the geometric similarity principle and the physical similarity principle, the key to which is to ensure the consistency of dynamic behavior in between the two [5,6]. Geometric similarity refers to that the subscale model has a similar aspect ratio and a similar way of connection to the shock absorber to the prototype test piece, while physical similarity is to ensure that the subscale model shows similar characteristics of vibration to the original one, with consideration primarily given to the similar natural frequency of subscale model to the original one.

For the vibration absorption System of aircraft transport, the mathematical model could be simplified into a SDOF spring-quality system with damping. The packaging used for aircraft transport is installed in the rear of the vibration table, which is connected to a model of it that is constructed based on the model of a single packaging box. As the the vibration table is purposed to simulate the vibration output of the bottom plate in the carriage, the mathematical model of the vibration absorption System for the packing box can be simplified into a SDOF spring-quality system with damping and classified as foundation vibration, as shown in Figure 1.

IOP Conf. Series: Materials Science and Engineering **727** (2020) 012018 doi:10.1088/1757-899X/727/1/012018

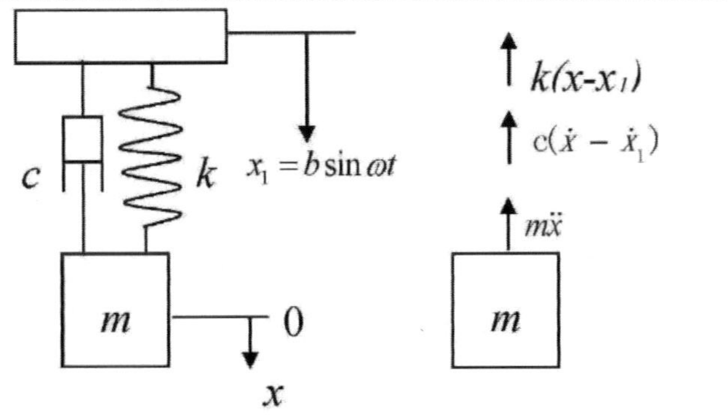

Figure 1. The vibration model of the packaging box.

The subscale model has a similar kinetic equation to the actual system, as a result of which the performance of the vibration absorption System obtained from the random test is sufficient to indicate that of the real system. Thus, a reasonable design of the parameters for the subscale model is essential, including mass denoted as m, rigidity denoted as k and damping denoted as c

- Simulation of the packaging box (m indicating mass)

As the packaging box is considered as rigid, consideration is primarily given to the mass and rigidity of the packaging box for the subscale model. With dimensions concerned, the connection to the vibration absorber needs to be considered in the first place, followed by the shape of the packaging box. Therefore, to achieve the purpose of simulation, a similar container to the packaging box can be used to be stuffed with what is easily changeable in mass.

- Selection of the shock absorber (k indicating rigidity and c indicating damping)

When the working characteristics of the vibration absorption System are identical to the real system, m and k can be obtained from the identical first-step natural frequency of the system, with the damping ratio unchanged. Otherwise, the interval with insignificant variation shall be taken wherever practical to perform test. The variation in response between the subscale model and the real model is calculated and rectified after experimentation.

4. Design of the testing and verification system for vibration isolation of aircraft transport

4.1. The overall plan

The testing and verification system is designed to validate the effectiveness of vibration isolation under the specified shock spectrum for the packaging box used for aircraft transport, and to provide guidance on how to optimize the vibration absorption System. Placed on the table of the verification system is a subscale model of the packaging box with a vibration absorption System. In the test, the bottom plate of carriage is subject to random vibration and then the output response is measured on the table. In doing so, the performance of the vibration absorption System is validated.

The rationale behind the verification system is shown in Figure 2. The signal generator is responsible for sending off the signal required by the verification system. The power amplifier is to amplify the current and voltage provided by the signal generator for the exciter, which provides the exciting force for the vibration table according to the input of the signal generator. The vibration table is the source of vibration for the system, that is to say, vibration stems from the table before the excitation spectrum is specified. The measurement and control system is purposed to measure the vibration value and exercise control of the exciter, so as to achieve the expected output of vibration load.

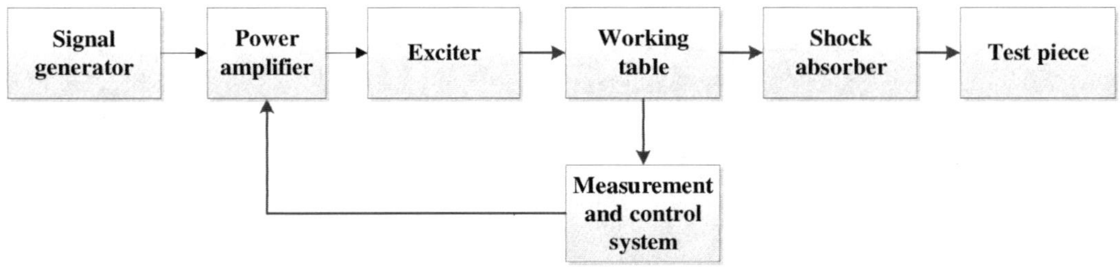

Figure 2. The flowchart of the verification system

4.2. The schematic diagram of the system

Figure 3 shows the schematic diagram of the verification system. In the subscale vibration absorption System, the air spring shock absorber with in-built dampers is taken as the component of vibration isolation. In order to improve the state of loading, the dampers are deployed at an angle in pairs, as shown in Figure 4.

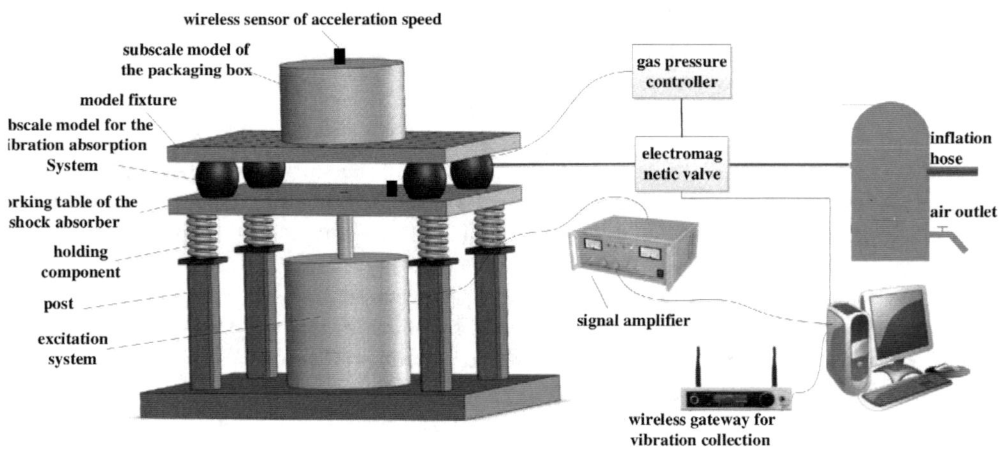

Figure 3. The schematic diagram of the verification system

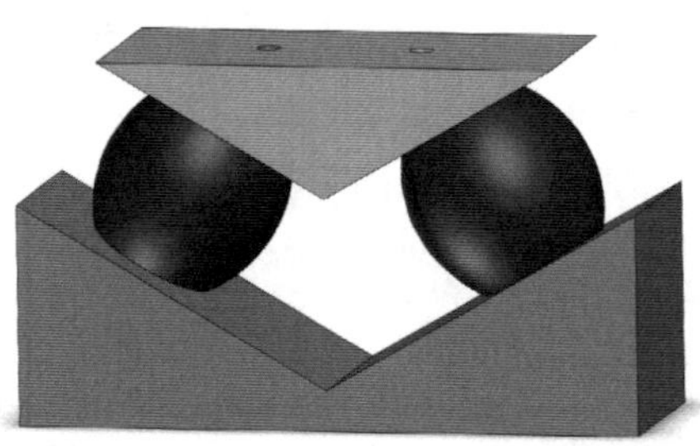

Figure 4. The schematic diagram of the air springs deployed at an angle in pair

4.3. Design of the excitation system

4.3.1. Excitation system. The excitation system represents the core equipment of the vibration test. In order for the occurrence of expected and controllable vibration that the test piece would be subject to, the excitation system is required to be capable of providing the alternating force with excellent waveform and sufficient amplitude and a certain steady force within the specified range of frequency. The former gives rise to acceleration speed for the test piece and the latter offsets the self-weight of it. In terms of power source, the excitation system can be either electrical, mechanical or hydraulic. The electrical excitation system is characterized by high frequency, excellent waveform, easy control and low capacity of single loading, which makes it suitable for the applications where the dynamic analysis and testing of components are required, such as aerospace, electrical instruments, auto parts and the likes. The hydraulic excitation system is characterized by ultra-low frequency, high excitation force and strong loading capacity, which makes it suitable for the applications where the reliability test has to be conducted on the heavy-weight and massive components or host equipment, such as shipbuilding, locomotive manufacture and car making. The mechanical excitation system is characterized by low frequency, high capacity, poor waveform and cheap affordability, which makes it suitable for the applications where the vibration resistance test is required, such as home appliance and instruments. Based on the above analysis, the electrical excitation system is chosen for the purpose of verification.

4.3.2. The main parameters of the vibration testing system. For an excitation system, the major parameters include the maximum thrust, the range of working frequency, the maximum load and the maximum acceleration speed.
The selection and use of an excitation system require a full consideration given to these parameters as referred to above. The relationship between thrust and acceleration speed is expressed as follows.

$$F = (m_1 + m_2)a \tag{1}$$

F represents the thrust generated by the vibration table, with the unit of N or kgf
m_1 indicates the mass of moving components, with the unit of kg
m_2 indicates the mass of the test piece and the fixture, with the unit of kg
a indicates the acceleration speed, with the unit of m/s^2 or g

The maximum acceleration speed of the excitation is based on no load. When load is applied, the maximum acceleration speed that can be achieved by the vibration table is less than when no load is applied. Besides, the extent of reduction relates to the load. Figure 5 shows the relationship between the thrust, load and acceleration speed.

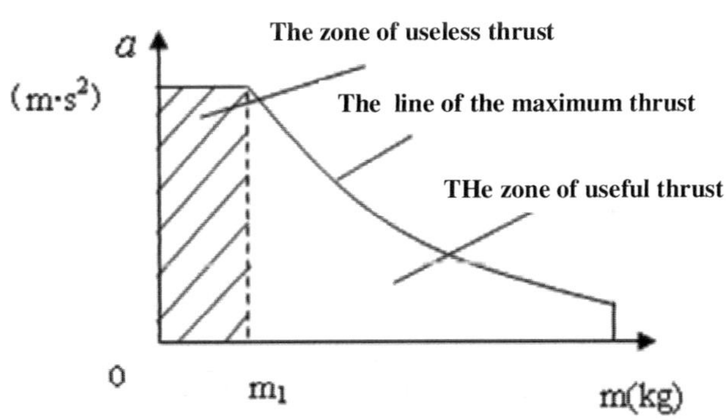

Figure 5. The relationship between the thrust, load and acceleration speed

In the case of sinusoidal vibration, the relationship between displacement, speed, as well as the amplitude value and frequency of acceleration speed is as follows.

$$V = A\omega = \frac{2\pi Af}{10^3} \quad (m/s) \tag{2}$$

$$a = A\omega^2 = \frac{A(2\pi f)^2}{10^3} = 2\pi fV \quad (m/s^2) \tag{3}$$

a means the amplitude of acceleration speed in m/s^2
A indicates the amplitude of displacement in mm
f denotes frequency in H_Z

It is a common practice that the multiple of gravity acceleration speed is taken to denote the acceleration speed. Therefore, equation 3 can be rewritten as:

IOP Conf. Series: Materials Science and Engineering **727** (2020) 012018 doi:10.1088/1757-899X/727/1/012018

$$a = \frac{A(2\pi f)^2}{9.8\times 10^3}\times 0.004Af^2\,(\mathrm{g}) \tag{4}$$

A set of parameters of DY-300 that are designed specifically for the low-frequency transport test are shown in Table 1.

Table 1. The parameter of the DY-300 electrical vibration table article.

Element	**Style**
Range of vibration frequency	2-2000 HZ
Rated sinusoidal thrust	2.94 KN
Rated random thrust	2.058rms
Maximum acceleration speed	367 m/s2
Maximum speed	1.50 m/s
Maximum displacement	40mm
Maximum load	130 kg
Mass of moving component	8 kg

Based on the parameters shown in Table 1, an analysis can be conducted of the basic performance of this excitation system. Further with this, it can be judged whether or not the requirements on package vibration test can be met. Meanwhile, equation 2 is used to calculate the maximum displacement and the crossover frequency at the maximum speed.

$$f_{AV} = \frac{V\times 10^3}{2\pi A} = \frac{1.5\times 10^3}{2\pi\times 20} = 11.9 \quad (\mathrm{H_Z}) \tag{5}$$

Equation 3 is used to calculate the maximum speed and the crossover frequency at the maximum speed.

$$f_{Va} = \frac{a}{2\pi V} = 38.9 \quad (\mathrm{H_Z}) \tag{6}$$

The DY-300 electrical vibration table can reach the maximum displacement at 11.9 HZ, the maximum speed at the range between 11.9 HZ and 38.9 HZ, and the maximum acceleration speed only at a frequency that is above 38.9 HZ.

An analysis is conducted of the impact of load on the output maximum acceleration speed of the vibration table. When the weight of the sample reaches the maximum load of 130kg, according to equation 3,

$$a_{\max} = \frac{F_{\max}}{m_1 + m_2} = \frac{2.94\times 10^3}{8+130} = 21.3(\ \mathrm{m/s^2}) \tag{7}$$

At this time, the maximum acceleration speed of the vibration table declines from 36.7 m/s^2 when there is no load to 21.3 m/s^2.

According to the applicable national standard, for tests conducted on cushion materials and packaging pieces, the vibration table is required to provide either fixed-frequency or swept-frequency vibration, with a fix acceleration of 0.5g and frequency ranging from 3 HZ to 100 HZ. The DY-300 electrical vibration table is capable to meet these requirements.

4.4. Design of the subscale vibration absorption System

4.4.1. Technical route. The technical route is shown in Figure 6.

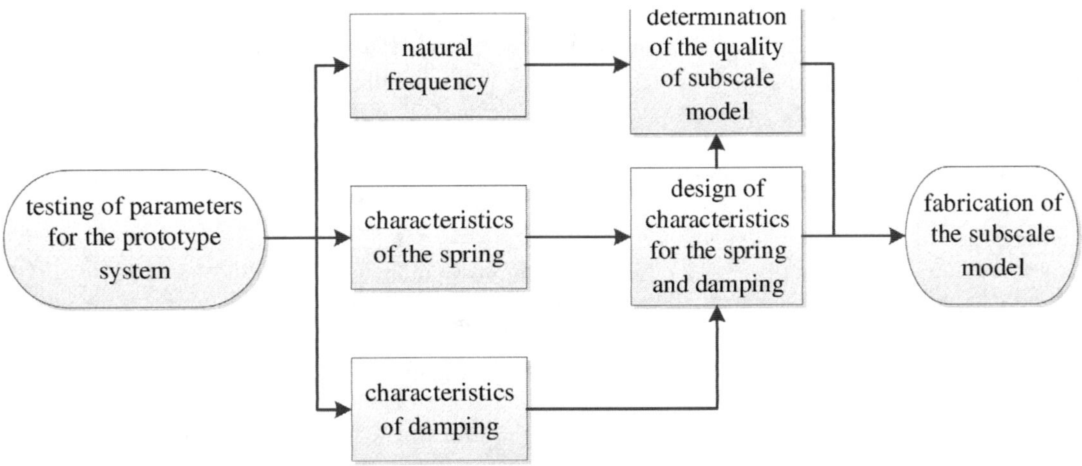

Figure 6. Technical route

4.4.2. The subscale model of the packaging box. The subscale model of the box is substituted with a rigid container, which can be filled with discrete object to adjust the overall mass.

4.4.3. Design of the subscale vibration absorption System. As the rigidity of the air spring is adjustable and significantly variable, a set of air springs can be applied to simulate the springs used in the vibration absorption System for packaging box in most cases. The air spring vibration absorption System consists of air springs, damper and rigidity-adjusting device. The air-adjusting device is controlled via a touch screen and is capable of making automatic adjustment of gas pressure based on the target value. Figure 7 shows the layout of the control panel.

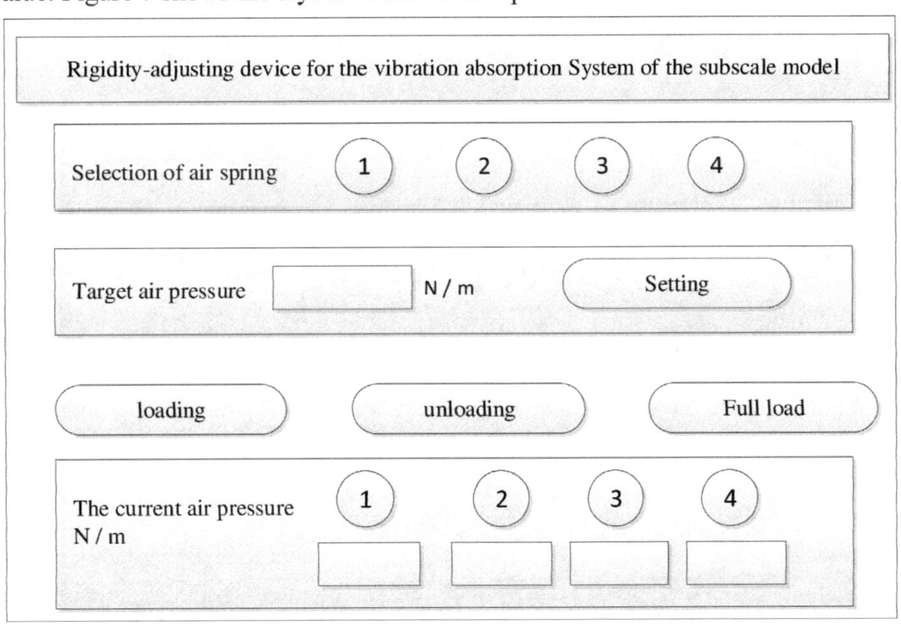

Figure 7. The layout of the control panel for the rigidity-adjusting device

4.4.4. Design of the data collection system and analysis system. A wireless data collection system is used to ensure convenience and reliability of testing on the relevant data. The JM5842 wireless testing device for acceleration speed is capable to meet the relevant requirements on testing.

IOP Conf. Series: Materials Science and Engineering **727** (2020) 012018 doi:10.1088/1757-899X/727/1/012018

5. The expected effect of the testing and verification system

Prior to transport, the quality of products to be transported and the conditions of transport are evaluated as input for testing and verification. The testing and verification system is applied to perform simulation of the vibration absorption System and identify the most appropriate layout of shock absorber deployment, which is beneficial to avoid the repeated running test that has to be conducted each time prior to transport, thus reducing labor and monetary costs.

6. Conclusion

By conducting a research into the approach to testing and verifying the vibration absorption System for the packaging of aerospace products, the simulation test table of vibration absorption System is designed and the vertical excitation that is similar to the bottom plate of carriage box is generated on the table. An environmental vibration test is conducted on the vibration absorption System, which provides a means to conduct effective evaluation of the system for energy efficiency and an approach to the optimization design of it.

Reference

[1] Yao Guonian. Experimental Conditions of the Accelerated Vibration Simulation for Special Product Transportation [J]. Journal of Xi'an Jiaotong University , 2009, 43(3): 74-77

[2] Lu Cailing. Vibration Test of Cargo Transportation [J]. Environmental Technology , 2011, 32(2): 13-17

[3] Ding Yong. The Research of Simulating Transportation Vibration Test Method based on Typical Road Spectrum [J]. Environmental Technology, 2012, 3: 49-52

[4] Zhu Xuewang. Application of Fatigue Damage Equivalence in Random Vibration Test[J]. Equipment Environmental Engineering , 2007, 4(1): 11-13

[5] Liu Zhanqing. Design and Validation of Small Proportion Model of Missile Engine's Flow Field [J]. Computer Simulation , 2008, 25(8): 51-54

[6] Gu Huafeng. Experimental and Finite Element Analysis of Heavy-Duty Automobile Disc Brake Noise [J]. Journal of Chongqing Institute of Technology , 2017, 31(4): 27-33

Factors Affecting the Accuracy of Estimation of BEV Cruising Range

Sun Zhicheng[1,2,*], Zhou Boya[1], Dai Tianlu [1], Meng Qingyu[1], Liu Zhichao[1], Xu Yaozong[1] and Li Su[2]

[1]China Automotive Technology&Research Center Co. Ltd.
[2]Hebei University of Technology, Tianjin,300300)

[*]Corresponding E-mail: sunshine649@163.com

Abstract. As the number of electric vehicles is increasing year by year, the accuracy of electric vehicle cruising range estimation in high and low temperature environments is one of the performances that consumers are very concerned about. In this paper, the accuracy of cruising range estimation is evaluated by the determination of coefficient, and the influence of ambient temperature on the estimation accuracy of cruising range is deeply studied. Eleven electric vehicles were selected to carry out the cruising range test at normal temperature, high temperature and low temperature respectively, and the influence law at ambient temperature of the accuracy of cruising range estimation was obtained. At the same time, the influence of the self-learning function on the accuracy of the cruising range estimation is also studied. Finally, the performance of the estimation of accuracy of the end mileage at different ambient temperatures is analyzed.

Keywords: environment temperature; electric vehicle ; range; range estimation accuracy; coefficient of determination

1. Introduction

China's new energy vehicle market has developed rapidly. By the end of June 2018, the number of new energy vehicles reached 1.99 million, of which 1.62 million were pure electric vehicles, accounting for 81.4% of the total number of new energy vehicles. Compared with fuel vehicles, electric vehicles have fast speed response, no pollution emissions, and low noise, but the virtual cruising range of electric vehicles is limited by battery capacity and power density. The cruising range, especially the cruising range at low temperature, is one of the performance indicators that electric vehicle consumers pay attention to[1].

Compared with traditional fuel vehicles, the range of electric vehicles is greatly affected by the ambient temperature. Studies have shown that temperature can affect battery life, mainly because ambient temperature can affect battery capacity. The capacity retention of the lithium battery at 0°C is 60% to 70%, 40% to 55% at 10°C, and only 20% to 40%at 20°C[2-3]. China has a vast territory, about 5,500 kilometers from north to south, and about 5,000 kilometers from east to west. Therefore, China has both a warm tropical climate and a plateau with snow all year round. Even if the temperature difference between different regions may become larger at the same season, it is important to study the

IOP Conf. Series: Materials Science and Engineering **727** (2020) 012019 doi:10.1088/1757-899X/727/1/012019

adaptability of vehicles at different temperatures.

During the actual use of electric vehicles, the meter shows that the remaining cruising range is another very important parameter, and the estimated accuracy of range can greatly reduce the driver's range anxiety[4]. However, the accurate or not range displayed on the dashboard, there is no unified evaluation standard at home and abroad. Therefore, it is necessary to carry out systematic research and establish an objective and feasible evaluation standard to evaluate it.

2. Experiment method

In this test, according to the "EV-TEST Management Rules" [5], the range test is divided into normal temperature range (25°C), high temperature range (35°C) and low temperature range(-7°C). Before performing the corresponding durability test, it must soak the vehicle in the corresponding environment for 12h to 36h, and then the NEDC condition test is conducted cycled. In the high temperature or low temperature durability test, the air conditioner needs to be turned on. In the high temperature range test, the internal temperature should be maintained between 23°C and 25°C. For low temperature range tests, the internal temperature should be maintained between 20°C and 22°C. The test of estimated cruising range accuracy and range test should be performed simultaneously.

2.1. Vehicle parameters

In this test, 8 vehicles with a body length of 4 m or more (conventional vehicle) and 3 vehicle bodies less than 4 m (mini vehicle) were selected. The vehicle parameters are shown in Table 1.

Table 1. Eleven types of vehicles parameters

Vehicle model	Curb quality (kg)	Range (km)	Battery capacity(kw•h)
1	1564	351	49
2	1710	320	48.3
3	1470	270	36
4	1410	301	38.5
5	1950	300	47.5
6	1598	300	41
7	2160	352	62
8	1610	200	30
9	1080	255	22
10	1175	151	29.2
11	855	151	18.2

2.2. Experimental procedure

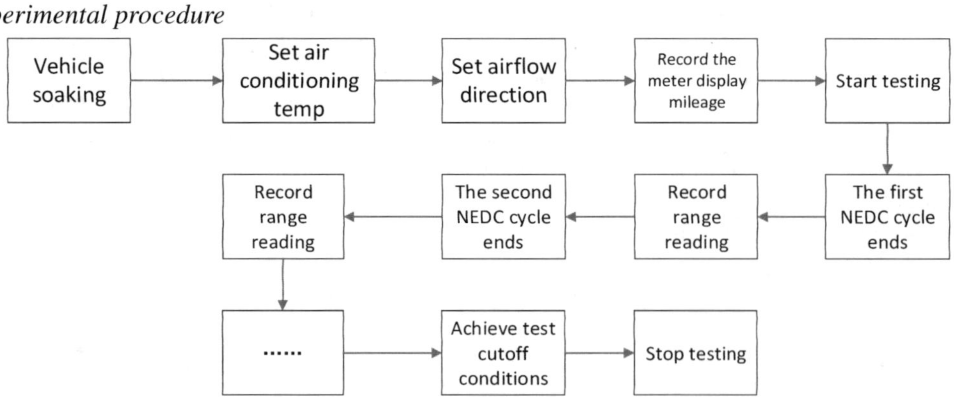

Figure 1. Test procedure.

IOP Conf. Series: Materials Science and Engineering **727** (2020) 012019 doi:10.1088/1757-899X/727/1/012019

The test process is shown in Figure 1. On the drum test rig, when the test cut-off condition is reached, the test under NEDC conditions is stopped[6]. When testing, record the remaining range displayed by the meter before the start of the test (data point of the 0th cycle) and the remaining range displayed by the meter at the end of each complete NEDC condition. When an NEDC condition is completed, a data is recorded, and after the test is completed, the actual remaining rangecorresponding to each the remaining rangeof meter is calculated based on the actual cruising range.

2.3. Determination of coefficient

In order to assess the accuracy of the actual remaining range display, it is necessary to convert it from a practical problem to a mathematical model to take the actual remaining range measured on the drum as a regression line and use the remaining range displayed on the dashboard as a dispersion point. In this way, the cruising range estimation accuracy problem is converted into the goodness of fit between the regression line and the scatter point. In this paper, the accuracy coefficient R^2 in statistics is used to evaluate the rangerange estimation accuracy.

Determination of coefficient: A numerical feature that represents the relationship between a random variable and multiple random variables. It is a statistical indicator used to reflect the reliability of the regression model's dependent variable [7].

The physical meaning of the Determination of coefficient is as shown in Figure 2. $\hat{y}_i$ indicates the actual remaining range, which is a regression line, and y_i indicates the remaining range displayed by the meter, which is the observed value.

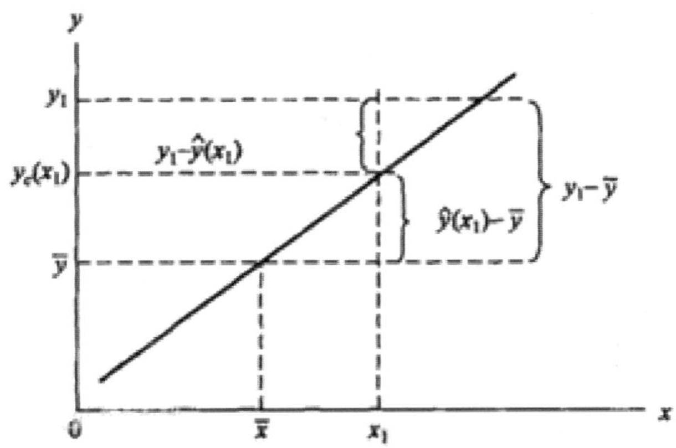

Figure 2. Variational decomposition diagram.

$$\text{Total sum of squares is: } SST = \sum(y_i - \bar{y})^2 \tag{1}$$

$$\text{Sum of residuals is: } SSE = \sum(y_i - \hat{y}_i)^2 \tag{2}$$

$$\text{Regression square sum is: } SSR = \sum(\hat{y}_i - \bar{y})^2 \tag{3}$$

The relationship between the three squares is:

$$SST = SSR + SSE \tag{4}$$

After derivation, the formula for determining the coefficient R^2 is:

$$R^2 = 1 - \frac{SSE}{SST} = 1 - \frac{\sum(y_i - \hat{y}_i)^2}{\sum(y_i - \bar{y})^2} \tag{5}$$

The closer R^2 is to 1, the larger the ratio of the sum of the squares of the regression to the total square sum, the closer the regression line is to each observation point, the better the fit of the regression line; on the contrary, the closer R^2 is to 0, the regression line The degree of fit is worse. When $R^2 \leq 0$, it is considered that the regression line seriously deviates from the observed value, that is, the degree of fitting is extremely poor.

IOP Conf. Series: Materials Science and Engineering **727** (2020) 012019 doi:10.1088/1757-899X/727/1/012019

3. Rangerange estimation accuracy

The remaining range and actual remaining range displayed by the meter of the electric vehicle are recorded in the experiment. The range displayed by the meter tested in the normal temperature environment is indicated by CB, and the actual remaining range is represented by CS. The range displayed by the meter tested in a high temperature environment is expressed in GB, and the actual remaining range is expressed in GS. The range displayed by the meter tested in a low temperature environment is indicated by DB, and the actual remaining range is represented by DS.

3.1. Test results of type some vehicles

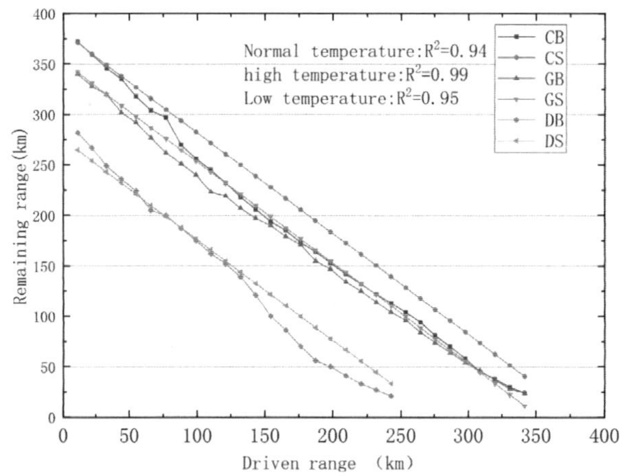

Figure 3. cruising range of No. 7 vehicle model at different temperatures.

Figure 3 shows the range test data of vehicle 10. It can be seen from the above three figures that difference the range displayed by the meter and the actual range are very small in different temperature environments. And the best performance occurs in the high temperature environment. R^2 is highest at high temperatures and lowest at low temperatures.

Figure 4 shows the results of the range experiment of the vehicle 7. It can be seen from the figure that the remaining range displayed by the meter in the vehicle 7 is not much different from the actual remaining range. In the low temperature environment, the difference between the remaining range displayed by the meter and the actual remaining range is large. R^2 is highest at normal temperature and lowest at low temperature.

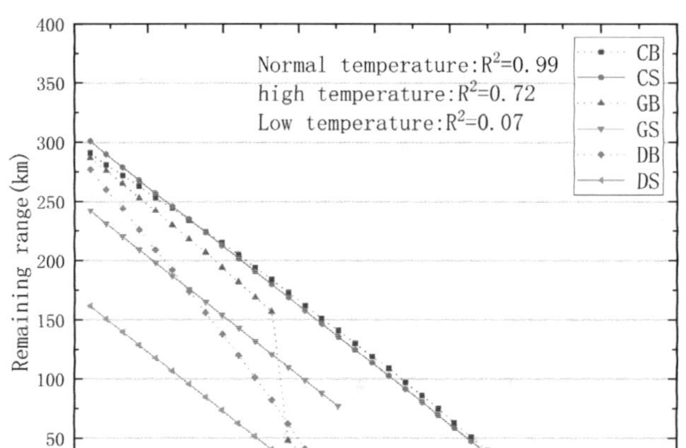

Figure 4. cruising range of No. 7 vehicle model at different temperatures.

3.2. Determination of coefficient of each vehicle type at different temperatures

Figure 5 shows the R^2 values of eleven types of vehicles at different temperatures. The average R^2 of the eleven types of vehicles is 0.88. Among them, the R^2 of eight types of vehicles is greater than 0.9 and the overall performance is good.

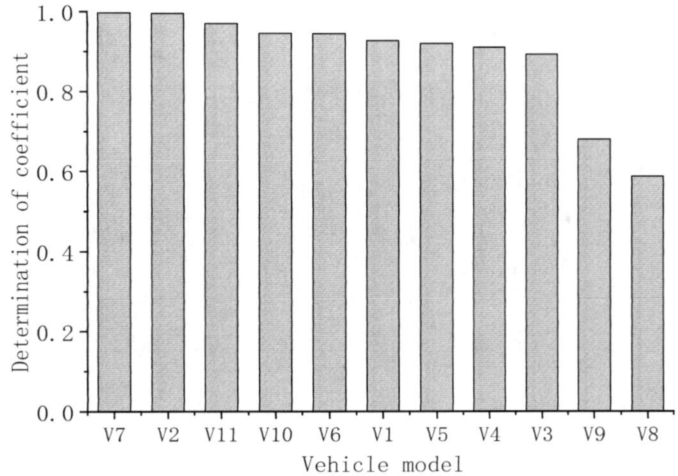

Figure 5. Comparison for R^2 of eleven type of vehicles under normal temperature.

The determination of coefficient of the vehicle 11 at high temperature is less than 0, so the cruising range estimation accuracy of the vehicle 11 in the high temperature environment is the worst. Figure 6 shows the Determination of coefficients of the remaining vehicles at high The average of R^2 for 10 types of vehicles is 0.81. Among them, only 5 types of vehicles have R^2 greater than 0.9, and vehicle performance begins to show a gap.

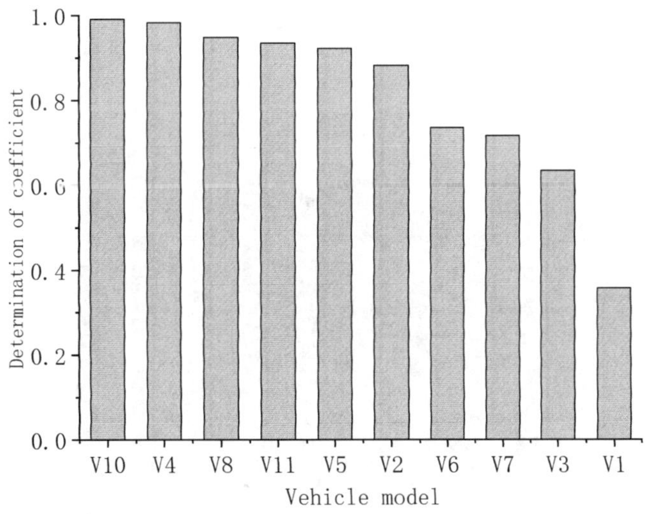

Figure 6. Comparison for R^2 ten types of vehicles under high temperature.

The Determination of coefficients of the vehicle 3, vehicle 5 and vehicle 9 in the low temperature environment are less than 0, indicating that the cruising range estimation accuracy of the three types of vehicles in the low temperature environment is very poor. Figure 7 shows the Determination of coefficients for the remaining eight vehicles in a low temperature environment. The average R2 of these 8 vehicles is 0.46, and onlyR2 of the only 2 vehicles is greater than 0.9, and the overall performance is poor.

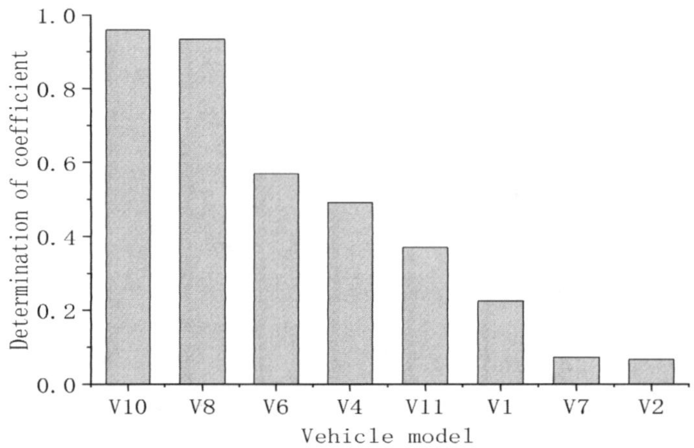

Figure 7. Comparison for R^2 of eight types of vehicle under low temperature.

3.3. The effect of self-learning on the Determination of coefficient

In order to improve the accuracy of cruising range estimation, some enterprises have added self-learning function in the control strategy, which can correct the current remaining range according to historical energy consumption, thereby improving the accuracy of the remaining range displayed by the meter [8-9]. In order to study the effect of self-learning function on the accuracy of cruising range estimation, this paper removes the data of the first two cycles by comparing and calculating, and collects data from the third cycle. The equivalent effect is to reserve a self-learning distance of approximately 22 km (approximately 11 km for an NEDC cycle) for the vehicle.

Figure 8 shows the ratio of R^2 between after the first two cycles and the total cycle of the eleven models in the normal temperature environment. As can be seen from the above figure, after removing the first two cycles in the normal temperature environment, R^2 becomes larger or smaller, and the average decreases by 5%. The accuracy of vehicle 10 andvehicle 11 varies greatly, R^2 Both fell by more than 24%.

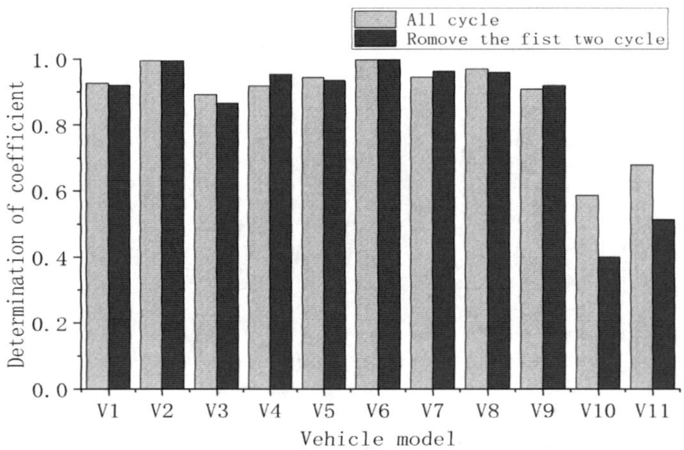

Figure 8. Change of eleven types of vehicles after removing the first two cycles under normal temperature.

Figure 9 shows the ratio of R^2 of the ten type vehicles between after the first two cycles removed and all cycles in the high temperature environment, the R^2 of the vehicle 11 is less than zero. The R^2 of the ten vehicles dropped by an average of 3.9%. Among them, R^2 of the types of vehicle 1 and vehicle 3 changed a lot, the accuracy dropped by more than 15%, and the others changed within 4%.

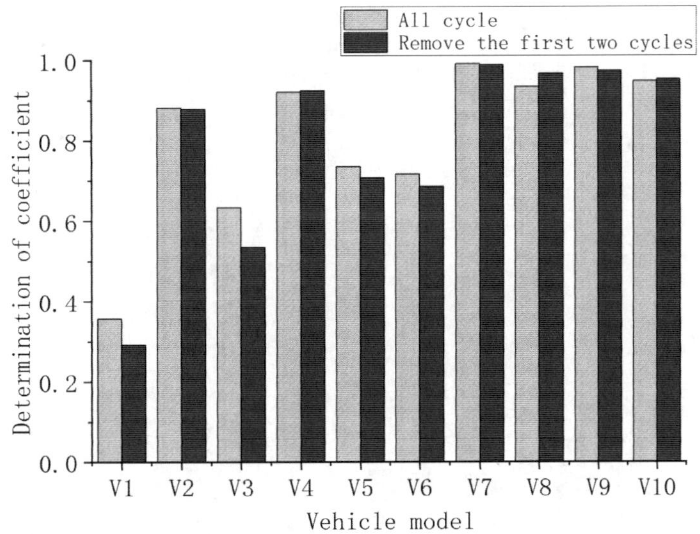

Figure 9. Change of ten types of vehicles after removing the first two cycles under high temperature.

Figure 10 is a plot of R^2 for the first two cycles and all cycles removed in a low temperature environment. Among the eleven models, the R^2 of the vehicle 2, vehicle 3, vehicle 4, and vehicle 11 is less than 0. It can be seen from the above figure that the determination of coefficient of the seven types of vehicle has averagly dropped to 18.7%. The variation of the 6 and 9 types of vehicle is larger, the R^2 is reduced by more than 40%, and the variation of others is within 6%.

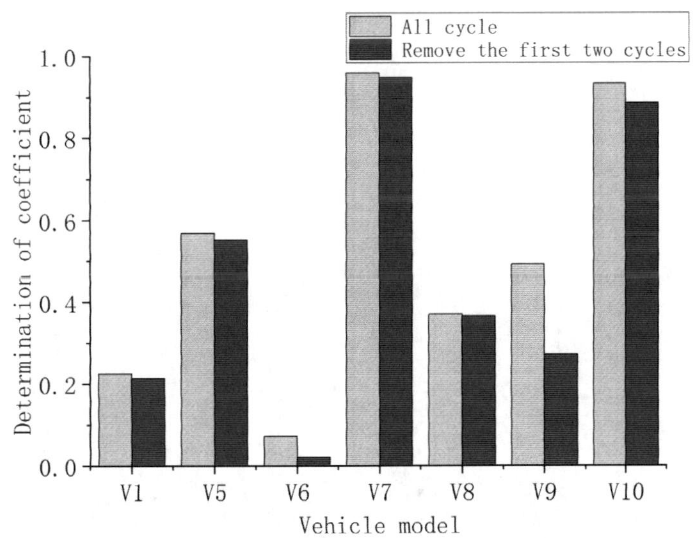

Figure 10. Change of seven types of vehicles after removing the first two cycles under low temperature.

It can be seen from the above data that the removal of the first two cycles has little effect on R^2 and does not change the overall result range. Therefore, whether the vehicle is equipped with a self-learning function has less influence on the result.

3.4 Estimated accuracy of the end range

When the battery power is low, the range anxiety of the electric vehicle driver is particularly prominent. This paper further studies the cruising range accuracy when the battery is low.

IOP Conf. Series: Materials Science and Engineering **727** (2020) 012019 doi:10.1088/1757-899X/727/1/012019

Figure 11 shows the Determination of coefficient R^2 of the last four cycles of vhhicle 7 at normal temperature, and the end range R^2 of the vehicle 5, vehicle 8, vehicle 9 and vehicle10 is less than zero. It can be seen from the above figure that the Determination of coefficient of the vehicle 7 has dropped by an average of 36.7%. Among them, the coefficient of determination of vehicle 2 dropped the least, down 20%, and the coefficient of determination of vehicle 7 dropped the most, down by 67.9%.

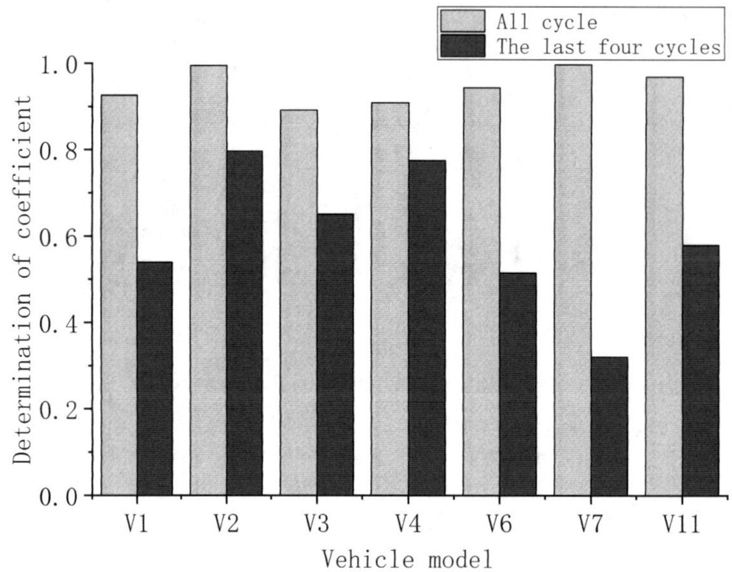

Figure 11. Determination of coefficient of the last four cycles in normal temperature.

Figure 12 shows the end range R^2 of vehicle 5 in a high temperature environment. The end range R^2 of other types of vehicle is less than zero. As can be seen from the above figure, the R2 of the 5 types of vehicle decreased by an average of 29.7%, of which the vehicle10 changed the most, the R^2decreased by 72.0%, thevehicle8 changed the least, and the R^2 increased by 0.37%.

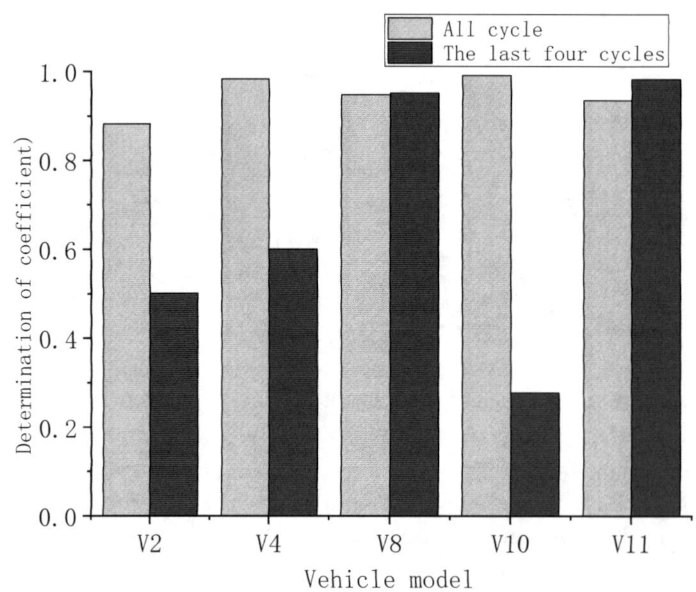

Figure 12. Determination of coefficient of the last four cycles in high temperature.

Figure 13 shows the end range R^2 of the two types of vehicle in a low temperature environment. The end range R^2 of others is less than zero. The R^2 of the two types of vehicle dropped by an average of 62.9%.

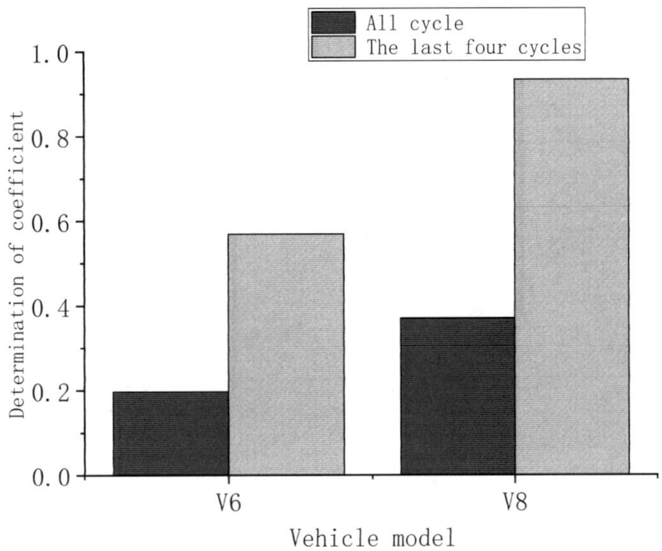

Figure 13. Determination of coefficient of the last four cycles in low temperature.

It can be seen from the above data that the R^2 of the end range of the electric vehicle is lower than that of the whole cycle R^2, and the R^2 in the low temperature environment is more decreased. Under low battery power conditions, vehicles generally do not provide accurate range estimates, which will increase the range anxiety of electric vehicle drivers.

4. Conclusion

The range estimation accuracy of electric vehicles at different ambient temperatures is also different. The normal temperature is the best, the high temperature is second, and the low temperature is the worst. In the normal temperature, the Determination of coefficient of nine types of vehicle reached 0.9, and the Determination of coefficient of only five types of vehicle in the high temperature reached 0.9, and the Determination of coefficient of only two types of vehicle in the low temperature reached 0.9.

It is small impact to the overall result that eliminating the influence of the self-learning function. So all data points can be directly collected when evaluating the accuracy of the cruising range estimation.

The end range estimate R^2 is greater than the R^2 of all cycles, and even some models have R^2 less than 0, which leaves a great space for the automotive industry to improve the accuracy of the end range estimation.

References

[1] Fu Lianxue. Research Report on Consumer Satisfaction of China's New Energy Vehicles in 2018[R]. Tianjin: China Automotive Technology Research Center Co., Ltd., 2018

[2] SongHongming,Liu Xiuqing,Xing Enfu,Zhang Sujiang.Study on Low Temperature Characteristics of Lithium Titanate Battery[J].Power Supply Technology,2018,42(06):818-820.

[3] S. Panchal, M. Mathew, I. Dincer, M. Agelin-Chaab, R. Fraser, M. Fowler,Thermal and electrical performance assessments of lithium-ion battery modules for an electric vehicle under actual drive cycles,Electric Power Systems Research,Volume 163, Part A,2018,Pages 18-27,ISSN 0378-7796,

[4] Jian Wu, Tong Li, Hao Zhang, Yanxiang Lei, GuangquanZhou,Research on Modeling and SOC Estimation of Lithium Iron Phosphate Battery at Low Temperature, Energy Procedia,Volume 152,2018,Pages 556-561,ISSN 1876-6102.

[5] EV-TESTManagement Rules (2017 Edition)

[6] Jia Junping. Statistics [M]. Beijing: Tsinghua University Press, 2004, 298-299,ISBN 7-302-08013-5

[7] Kai Liu, Jiangbo Wang, Toshiyuki Yamamoto, Takayuki Morikawa,Exploring the interactive effects of ambient temperature and vehicle auxiliary loads on electric vehicle energy consumption,Applied Energy,Volume 227,2018,Pages 324-331,ISSN 0306-2619.

[8] Christian Fleischer, WladislawWaag, Ziou Bai, Dirk Uwe Sauer,On-line self-learning time forward voltage prognosis for lithium-ion batteries using adaptive neuro-fuzzy inference system,Journal of Power Sources,Volume 243,2013,Pages 728-749,ISSN 0378-7753,

[9] Claudio Sbarufatti, Matteo Corbetta, Marco Giglio, Francesco Cadini, Adaptive prognosis of lithium-ion batteries based on the combination of particle filters and radial basis function neural networks,Journal of Power Sources,Volume 344,2017,Pages 128-140,ISSN 0378-7753.

AUTHOR INDEX